4-1-97

Fundamentals of manufacturing for engineers

Fundamentals of manufacturing for engineers

T. F. Waters

UCL PRESS

© T. F. Waters, 1996

First published in 1996 by UCL Press

UCL Press Limited
University College London
Gower Street
London WC1E 6BT

and

1900 Frost Road, Suite 101
Bristol
Pennsylvania 19007-1598

The name of University College London (UCL) is a registered trade mark used by UCL Press
with the consent of the owner.

British Library Cataloguing-in-Publication Data
A CIP catalogue record for this book is available from the British Library.

Library of Congress Cataloging-in-Publication Data are available

ISBN: 1-85728-338-4 PB

Typeset in Times and Gill Sans.
Printed and bound by
Bell & Bain Ltd, Glasgow.

Contents

v

Preface

If you are examining this book for the first time, you probably have an interest in manufacturing, and while there are numerous texts in this topic area this book offers a very different approach.

All areas of higher education have been subject to radical changes in recent years, most of which have had a profound effect on both what is taught and how it is taught within the time available. Science-based courses have also had to adjust to accommodate students enrolling with entry qualifications in subjects not ideally suited to the course they wish to study. Engineering has been particularly affected by this, with many students, for example, possessing only modest mathematical proficiency.

In an attempt to minimize the adverse effects of limited mathematical skills during the early stages of the course, I have kept "hard sums" to a minimum, and only employed mathematics where it is essential to explain a fundamental principle.

Most courses have changed dramatically in both content and emphasis, the trend being towards a much larger number of topics being taught, but as total study time has not increased (it has even reduced in many cases) depth of study has had to be sacrificed. Furthermore, with the introduction of the two-semester system on many MEng/BEng degree and diploma courses, coupled with vastly increased class sizes, the lecture and tutorial time available to each student has been cut to the bone. This puts greater onus upon the student for self-study, which, in turn, demands textbooks written in a style that is both interesting and as easy to read as possible. This book has been written with these objectives in mind, particularly as it is aimed principally at first and second year students.

With such a descriptive subject as manufacturing, and with so many students having little or no previous practical experience, if a book is to become the adopted text in this field it is important that it is structured and written in such a way that it assumes no prior knowledge in any given topic and, wherever possible, relates new concepts to familiar situations encountered in everyday life. I have tried to achieve this, but to avoid continually interrupting the natural flow of a topic I have employed a system of Note boxes. I have used a similar idea to provide small participation exercises at various points in the text where I feel they would aid the grasping of the fundamentals involved.

This book, as the title suggests, is not intended to be the ultimate comprehensive tome covering every conceivable manufacturing process in considerable detail. Books following this approach are usually so large as to be very expensive and tend to overwhelm most first year students. While this book covers almost all of the processes regularly encountered in industry, by omitting a few specialist and rare processes it has been possible to present the basic essentials of manufacturing that all engineering students should grasp by the time they begin their industrial career. This does not mean that the treatment of topics has been, in any way, trivialized, but rather essential principles are not clouded with a mass of non-essential detail.

It is generally accepted that a simple drawing or photograph is worth a thousand words, and for that reason this book relies heavily upon illustrations – well over 300 in fact. It was in the acquisition of many of them that I was most heartened by the willingness and enthusiasm with which industrial companies throughout the European Union supported me, often at considerable expense to themselves. I am most grateful not only for their diagrammatic contributions, but also to their engineers for assisting, making recommendations and appraising the technical content in the light of modern manufacturing practice in their areas of expertise.

Very special thanks must also go to Roger Watson and Dr Jack of Sandvik Coromant, David Coleman and Drs Brian Powell and Mike Asteris of Portsmouth University, John Hyde of Kryle Machine Tools International, David Whittaker of D. Whittaker & Associates, Dr Tim Weedon of Lumonics, Ken Richards of the Welding Institute, Geoff Duckett of William Cook, Peter Beim of Platarg Engineering, Drs Frank Wilson and Ken Lewis of British Steel, Dr Terry Lester and Steve Skidmore of Metallisation, Kevin Nixon of Alumasc, Ken Edmett of Mitutoyo (UK) and Peter Dunn of Tesa Metrology. Without the technical support and encouragement of these and other highly regarded engineers, this book would not be so technically up to date.

Wherever possible I have based the case studies provided at the end of Chapters 2–11 on real industrial problems. Because of this, while it is essential to draw upon material contained in the chapter concerned, many other facets of engineering are also deliberately included. *These case studies are meant to be further learning exercises*, and students should therefore not be too concerned if they feel that they would probably not have solved the problems cited purely by reading the relevant chapter. They should appreciate that this is not what happens in real life, and that knowledge in many aspects of engineering is invariably called for to solve most industrial problems. Such broad experience takes much time to acquire, but one always learns a lot from experienced colleagues!

Learning can only be fun if it is an enjoyable experience, and this involves the use of easy-to-read books as well as the student feeling that tutors are genuinely knowledgeable and enthusiastic about their subject. I can do nothing to influence tutor motivation, but if this book helps to pass on only some of the broad experience that I have gained by 20 years in industry followed by a further 20 years as a university lecturer and industrial consultant then the effort required to produce it will have been worthwhile.

Finally, if any student (or lecturer) feels that subsequent editions of this book can be improved to make the learning experience even more effective and enjoyable, I would welcome, via the publisher, such suggestions.

T. F. Waters, MSc (Eng), DIC, CEng, FIMechE

This publication has been sponsored by

Chapter One
Engineering manufacture in modern society

All too frequently students skip reading the introductory chapter of textbooks such as this because they do not see them as containing the essence of the basic manufacturing processes that they are required to learn, and hence they believe that they are unlikely to contain potential examination material! The fact that you are reading this now is encouraging, so please continue to read the remainder of this chapter before embarking upon the more detailed studies contained in subsequent chapters.

In addition to providing the reader with a broad overview of current engineering manufacture, and how it is changing to meet the challenges of the twenty-first century, this first chapter also briefly describes how manufacturing fits into the overall economy of a country, and how it contributes to the material standard of living of its population. The important issues of safety and the need to preserve our environment are also considered.

1.1 Engineering in a modern industrial economy and as a career

If a country is blessed with huge natural resources of oil and minerals, ignoring political issues, it should be possible to provide a high standard of living for its people. This can be achieved either by buying in all the products and services required or by establishing a solid industrial base of its own. Certain Middle Eastern oil-producing countries are good examples of the former approach, and the United States is a prime example of the latter.

Unfortunately, natural reserves do not last forever, so countries that fail to establish an industrially based economy, although still able to pay for all the goods and services that they want for a limited period, will eventually be faced with a rapidly declining standard of living as their natural wealth reserves become exhausted. However, in countries that can offer a wide range of goods and services that others wish to buy reliance upon natural wealth is reduced, because such an industrially based economy is inherently wealth creating. This is because any commercially successful enterprise must make a profit, i.e. its income must exceed its costs, and is therefore a wealth creator rather than a wealth user.

Engineers are an essential element in such a wealth-generating economy, and the more innovative their designs, and the more efficiently they make their products, the greater will be the level of wealth generation. Good examples of countries that have achieved such success in recent years are Germany and Japan. Indeed, manufacturing has been so successful in improving productivity per employee in the past two decades that it is now possible to meet the increasing demand for goods using fewer and fewer workers. This presents those working in manufacturing with ever greater technical challenges, particularly as an increasing number of

Third World countries enter the world's industrial market place. Figure 1.1 illustrates clearly how heavily industrialized countries are far more productive per head of population (per capita) than less industrially efficient, labour-intensive ones. But for how long will countries such as India and China be content to generate just 1 per cent of the gross domestic product (GDP) per capita that Japan currently achieves?

So if you wish to pursue a challenging career that offers the opportunity to make a positive contribution to the future prosperity of your country, as well as providing a high degree of personal satisfaction, then engineering is probably a good choice.

Note   The key to efficient production is the concept of *added value* (§10.5). The more effectively that the value of an item is increased at each stage of manufacture, the more productive and profitable the overall production system is likely to be. Even service industries such as hospitals and schools are turning to this method of assessing their operational efficiency, although the parameters used in measuring added value are less obvious than in a manufacturing environment, and are often the subject of heated debate.

1.2 Technical and organizational considerations

Engineering manufacturers are extremely diverse, ranging from drawing pin manufacturers to producers of the latest commercial jet aircraft. While the difference in technology and manu-

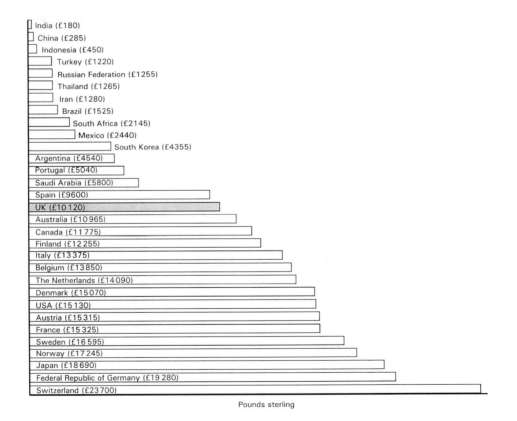

Pounds sterling

Figure 1.1   GDP per capita in 1992 for countries with a GDP in excess of £50 000 million (World Bank 1994).

facturing complexity between two such extreme examples is enormous, the commercial basis of operation is similar because, whatever the end product, it has to be made in some way, and there must be a customer base. Therefore a sales/marketing organization is needed to locate potential customers, and an efficient manufacturing system is required to turn raw materials into the finished product. This is of course, an overly simplistic description of any manufacturing company's structure as it ignores many essential supporting activities such as finance, personnel, secretarial support, transport, building services maintenance, etc. However, most engineering manufacturers have a broadly similar structure, and this usually involves a product flow system typical of that shown in Figure 1.2. Because Figure 1.2 illustrates a typical routing for an order to produce items in a company's existing range, design and production planning activities have been omitted.

1.2.1 The essential relationship between design and manufacture

Every company must continually develop new and improved products, as it is a fact of life that no company can commercially stand still. It either grows or dies.

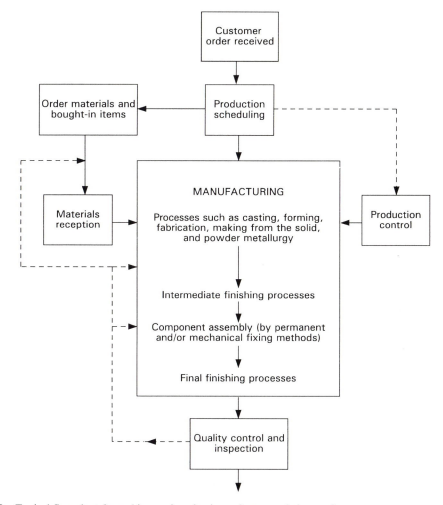

Figure 1.2 Typical flow chart for making engineering items from an existing product range.

Because the sales and marketing department is best placed to know the needs of the market place and what competitor companies are offering, it must play a major role in helping to draw up future product specifications. Nevertheless, the technical responsibility for producing the detailed specifications required to make these new products lies with the design function.

Unfortunately, in the past, designers have tended to consider little, if at all, the implications of their designs on manufacturing costs. This was mainly because of a lack of manufacturing process knowledge on the part of the designer, but happily this problem is now being overcome as an increasing number of engineering courses include "Design for Manufacture" as a core theme.

Because manufacturing costs are usually at the heart of product selling price calculations, it is vital that designers and production planning engineers work as equal partners in a team to produce the most cost-effective designs. With this approach design engineers soon realize, for example, that calling for, say, six different radii on a particular part where two would have been acceptable greatly increases the number of cutting tools required, or that demanding unnecessarily close component dimensional tolerances will probably result in the need for much more expensive manufacturing processes.

Note   A production planning engineer is someone who plans in detail how a part can be most efficiently made, taking into account the number required, the materials involved, and the equipment available. It is also usually this person's responsibility to specify and/or design any special fixtures and tooling that may be needed, as well as to write any computer-controlled machine tool programs that may be required.

1.2.2 Some modern approaches to efficient manufacture

In addition to closer links being forged between the design and manufacturing functions, other advances have been made in the quest to increase production efficiency and to minimize the total time required for a new product idea to reach the market place – an essential requirement as product life cycles become progressively shorter.

Another major change that has taken place in recent years is that the customer now demands much greater product choice than ever before. This is particularly so in the automotive industry. Gone are the days when Henry Ford declared that customers could have any colour they liked as long as it was black. The current situation is very different and is well illustrated by an advertisement from the car manufacturer Volvo (Fig. 1.3) – and this applies to just one of their models! Such increasingly demanding marketing requirements have resulted in more integrated and flexible systems of design and manufacture being adopted, three such techniques being concurrent engineering, just-in-time and computer-integrated manufacture.

1.2.2.1 Concurrent engineering

If one examines the traditional way in which a new product is progressed from the conceptual design stage to quantity production, it will be seen to follow a logical number of sequential steps (Fig. 1.4a). What is not so apparent until one has lived through such exercises is that an enormous number of modifications, redesigns, revisions and up-dates are usually generated, mainly as a result of problems encountered at the manufacturing stage. A figure quoted in 1992 (Gould 1992) suggested that in the UK 50 per cent of all design office time is spent on design modifications, but in Japan this figure is nearer 10 per cent. It is claimed that the main reason for Japan's greater efficiency is that manufacturing requirements and their implications for design are resolved long before production starts, i.e. concurrently with the initial design and development stages (Fig. 1.4b).

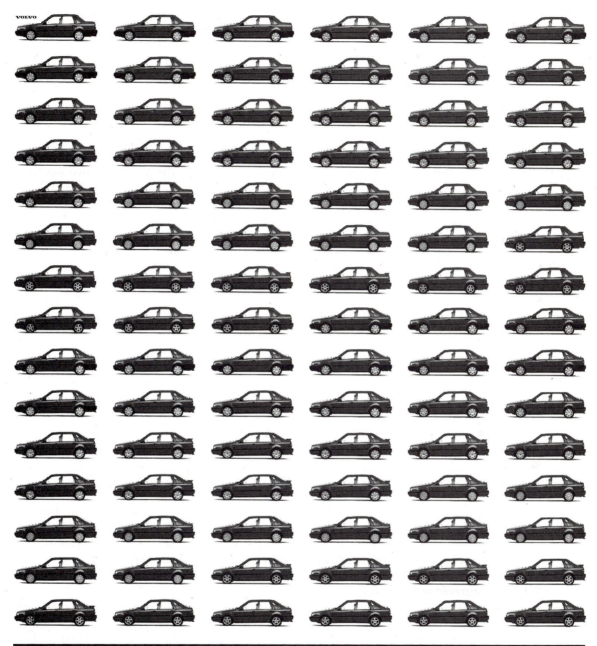

THEY'RE ALL DIFFERENT.

No. Your eyes don't deceive you.

Not one of these cars from the Volvo 400 series is the same.

Upon closer inspection you'll find variations in engine size, specification or body type.

But that shouldn't really surprise you. Volvo offers you a wider choice than any other manufacturer in this sector.

Which means that you decide exactly where your money goes.

If you'd like a 1.6 engine with a luxury package there's a Volvo for you two along in the eighth row.

And if you're looking for a 1.9 turbo diesel engine with sports package go for the fifth Volvo along in row eleven.

Figure 1.3 Advertisement for ninety variants of just one car model. (Courtesy of Volvo Cars Ltd, Marlow, Bucks.)

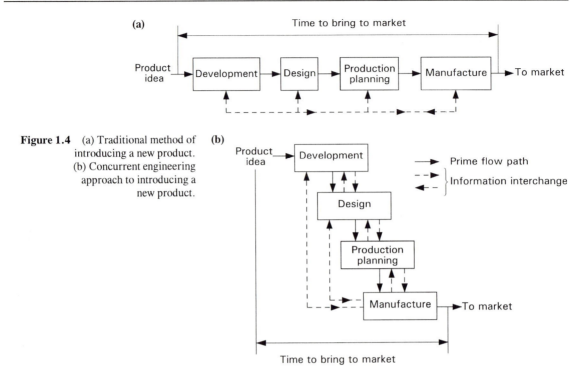

Figure 1.4 (a) Traditional method of introducing a new product. (b) Concurrent engineering approach to introducing a new product.

The subject of concurrent engineering is complex (Lambert 1994), and makes extensive use of modern computer technology, but it is sufficient here for the reader to understand the basic philosophy, which is to work simultaneously on as many stages of development, design and production planning as possible. This then solves many problems before they occur, as the product design and the optimum method of manufacture are developed concurrently.

1.2.2.2 Just-in-time manufacture

When a batch of components is ordered, traditional methods of manufacture have led to the attitude that, because a few problems are bound to occur somewhere, an element of rework or scrap replacement is unavoidable. This means that one accepts either that late delivery of part of the batch is inevitable or that a slightly larger batch than is required must be ordered to overcome the problem, i.e. the just-in-case strategy. Such a negative approach is, of course, a very efficient way of losing money, and begs the question of why a few items will always be made incorrectly. How this problem may be overcome is briefly discussed in Chapter 10 (§10.4), but here we are concerned with just-in-time (JIT) manufacture, rather than the wasteful just-in-case approach.

A basic principle of JIT (Bicheno 1991) is that goods are made not only in exactly the number required, but also at the last possible moment, while still meeting the required delivery schedule. An implication of this strategy is that each part must be made right every time, and this applies equally to all bought-in items, which must also be delivered on time without fail. This requires great faith in one's suppliers who, in turn, must therefore also operate on the JIT principle.

Because of the large and continuous volumes of production common in the automotive industry, car manufacturers usually have most to gain from operating JIT. Indeed, JIT was developed by Toyota in Japan. Apart from ensuring that the highest levels of manufacturing consistency are maintained, JIT also minimizes stock levels of bought-in items that have to be stored.

A classic example of the use of JIT occurs at the Nissan car plant in Sunderland (Wickham 1993), where seats are ordered for the Micra model just 45 minutes before they are required on the assembly line. This is achieved by the seat supplier being located 3 km away, and being in continuous computer contact. Five other local suppliers also supply parts on a similar JIT basis to this production plant.

A useful commercial benefit of JIT is that one sometimes receives payment for the finished product before component suppliers have to be paid, thus establishing a very favourable cash flow situation.

1.2.2.3 Computer-integrated manufacture

With the increasing use of computer numerically controlled (CNC) machine tools, and the development of sophisticated computer software packages designed to carry out administrative functions such as production scheduling and control, automatic materials ordering, etc., numerous attempts have been made to marry up these activities, and so create as automated a factory as possible. This is usually referred to as computer-integrated manufacturing (CIM), and is yet another attempt to minimize the time taken to bring a new product to the market place.

While considerable success has been achieved in certain specialist areas, there are problems, particularly in getting the various constituent parts of the system to interact fully with one another. This is largely because there are no standardized interface protocols between different computers and their related software, and attempts to find a universal solution to this basic problem have so far met with only limited success (Kief & Waters 1992). Nevertheless this is clearly the correct philosophy for the future, and increasingly industry will continue down this route.

One area at the heart of any CIM system is the linking of computer-aided design (CAD) with computer-aided manufacturing (CAM). CAD involves the use of a computer-based drawing system, which generates drawings and then stores them in digital form on the computer's hard disk. This technology has now largely replaced the traditional drafting process of using a drawing board and pencil. As engineering drawings are the basis of all subsequent manufacturing operations (Fig. 1.5a), and as CNC machines require computer programs to produce the required parts, it is logical that a common database be used (Fig. 1.5b). It should then be possible to go directly from a CAD drawing to the CNC machining programs with minimal human intervention.

To date most CAD/CAM systems still require a considerable degree of input from production planning engineers, but there are a number of CAD/CAM systems that can make components directly from CAD drawings, without human involvement.

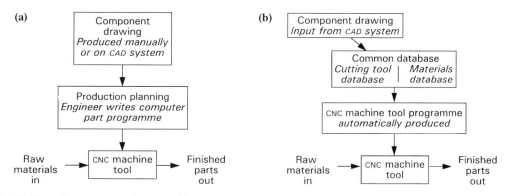

Figure 1.5 (a) Manually programmed CNC machine tool. (b) Integrated CAD/CAM system.

1.2.3 The importance of safety

Most activities in life, from making a cup of tea to crossing the road, can be dangerous. However, we all carry out such actions daily without mishap because we act in a manner that takes account of the risks involved. The same principle applies, of course, to activities at work.

Engineers, like everyone else, have a duty of care (Factories Act 1961) towards staff in their charge to ensure that any equipment that is potentially hazardous is suitably guarded, and that the staff operating such equipment are properly trained in its use. Furthermore, there is also a legal responsibility to ensure that equipment is designed and made so that it presents no hazard to anyone likely to come into contact with it. Maintenance should be as simple as possible: many accidents are caused by equipment that has been poorly maintained, often because of the complexity of carrying out such work.

Safe working practices are essential, and the attitude of prevention is better than cure should be encouraged at all times.

1.2.4 Environmental considerations

People are becoming increasingly concerned about damage to the environment and about the rate at which the world's natural resources are being consumed. There is ample evidence to justify such worries, and one of our prime considerations as engineers should therefore be to minimize the use of materials and non-replaceable energy resources.

Certain manufacturing processes are inherently less environmentally friendly and more expensive than others. It therefore makes economic sense to select processes with both of these factors in mind. Examples of energy-hungry manufacturing processes are the making of parts from solid material when a large proportion of the original volume of material must be machined away, the use of metals where composites will do the job equally well, and casting parts that can be made equally well by less polluting and less energy-demanding processes. Thus, the nearer the initial shape and form of the raw material is to the final product, the more economic and environmentally friendly will generally be the method of manufacture.

One method of minimizing the use of virgin material is to adopt a policy of recycling materials whenever possible. This also makes commercial sense, as pollution legislation is likely to become increasingly severe, and the cost of all forms of energy is likely to increase.

On a less global scale, employee environment is of great importance to the wellbeing and working efficiency of staff, whether they be computer operators or car assembly line workers. In addition to the safety aspects referred to in §1.2.3, good lighting, minimal noise levels and pleasant atmospheric conditions are all important factors, as is the efficient ergonomic design of the work place.

Exercise Ergonomics is the study of the relationship between workers and their environment. To see just how important this can be, type the contents of this exercise using either a typewriter or a personal computer while in a standing position.

Time spent ensuring that working conditions are optimized can result in an unexpectedly large increase in productivity, and usually does not involve large expenditure. It also helps to reduce accidents, particularly in manual working areas.

1.3 Economic factors

1.3.1 What constitutes demand?

The only way that a manufacturer will sell its products is by convincing potential customers that they are superior to the competition in one or more of the following areas: price, design excellence, quality of manufacture, availability and product reliability supported by an efficient after-sales service. There are other factors such as appearance, perceived value for money in its own right and in relation to the competition, brand name, etc., but they are all indirectly linked to the five primary factors. These are factors that influence a customer who is actively shopping, i.e. is looking for a specific item to meet a known need, and consideration of these influences is termed consumer-led marketing.

In contrast, technology-led marketing, while still having to demonstrate value for money, quality and so on, has the added task of persuading potential customers that they really need the new product on offer. This is much harder of course, because it is often difficult for potential customers to jump the technology gap that may be involved. Notwithstanding this, if design and production staff have performed their tasks well, the rewards can be great. This is particularly true if, as is often the case, one has patent protection for a number of years following launch.

A good example of such a technology-led product is the touch trigger probe (Chs 9 and 11), which has revolutionized the way that precision measurement is now performed in engineering manufacture.

Exercise   Make a list of a few consumer-led and technology-led consumer products that have been launched in recent years.   *Hint*   Sir Clive Sinclair
How many of your list of technology-led products were commercially successful?

1.3.2 Investment in people and equipment

The most valuable assets of any commercial company are its customer base and its staff. Equipment, buildings, etc. are all secondary as they are replaceable and, other things being equal, are mainly a question of money. Engineers are usually thorough in their technical assessment of equipment before purchase to satisfy themselves that it will do the job when installed. Costs associated with the purchase and installation of such equipment are relatively easy to establish, but this is not the case when it comes to employing staff.

It takes time and a great deal of effort to find and train the right person for a given job. Once in post not only is it expensive to repeat the recruitment exercise should the incumbent decided to leave, but untold expense and disruption can be incurred in filling the post temporarily with staff not specifically trained for that job, or by sharing the work out among several staff as a temporary addition to their normal duties.

It therefore makes sense to apply adequate resources to support the recruitment, training, long-term career development and welfare of staff. Sadly, this is not always the case, and engineers in particular sometimes fail to appreciate that the acquisition of the most suitable and well motivated staff is even more important than buying the most appropriate equipment. Perhaps this is because the assessment of people is more difficult than equipment evaluation, and most engineers receive insufficient formal training in staff recruitment and management techniques. This is unfortunate, as many smaller companies do not have personnel departments, and engineers therefore find themselves responsible for selecting their own staff.

1.3.3 Money matters

It is a sad fact that many companies continue to become insolvent because of poor financial management and, while engineers are not expected to possess in-depth accounting knowledge, a basic appreciation of the fundamentals is essential. Fortunately, most engineering courses now provide such training, and this is helping to eliminate the long-standing reputation of engineers and scientists as poor business managers. It is vital to appreciate, for example, the difference between capital and revenue budgets, and the importance of efficient cash flow control – one of the most common causes of company bankruptcy – as well as being able to read and understand balance sheets and profit/loss accounts.

1.4 Mastering the basic processes and terminology used in engineering manufacture

Whatever career one chooses, before being in a position to make a significant contribution, it is always necessary first to become familiar with the terminology and basic principles relevant to that field of work. This can be boring at times, but unlike the medical profession at least engineers do not have to learn any Latin terminology! This book makes every effort to make the learning experience as interesting and relevant to day-to-day living as possible, but unfortunately this does not relieve the student from the task of studying in detail the processes described it the remaining chapters of this book and committing the basic principles to memory.

The learning process is often found to be more meaningful if the reader has an initial picture of how each of the topics to be studied fits into the overall "package of knowledge" that must be learned (Fig. 1.6). It also helps reinforce the concept that, while learning must be presented in digestible lumps, once training is complete it is necessary to be able to draw upon the whole pool of knowledge because so many topics are inter-related. This is equally true across subject boundaries. For example, in addition to an in-depth knowledge of manufacturing matters, pro-

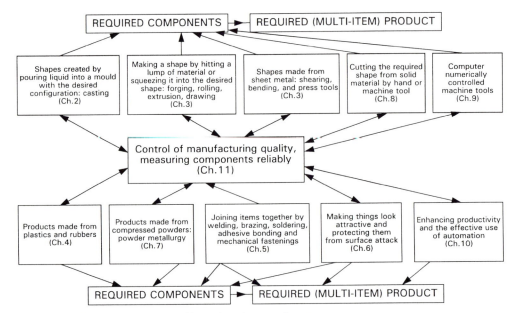

Figure 1.6 Key topics and processes encountered in engineering manufacture.

duction engineers, to be truly effective, must also possess a working knowledge of design, materials science, mathematics and other related subjects.

Most engineering products are made by assembling a number of separate items and, although the efficiency with which assembly is carried out is vitally important, this book tends to concentrate mainly on the processes involved in producing individual parts.

Manufacturing usually involves making items from liquid or solid raw material by either controlled manipulation and/or material removal processes. Changing material properties to enhance component performance and durability is also frequently involved, although this is normally the province of the materials scientist.

The surface finish of both discrete components and complete products is crucial as the customer's first impressions are very important; the product's durability is also critical. A chapter is therefore devoted to this frequently neglected area (Ch. 6).

Last, but by no means least, is the matter of quality assurance. It is essential that any product fully conforms to specification, in both performance and reliability, and as this crucial topic impinges on all areas of manufacturing (Fig. 1.6) the final chapter of this book is dedicated to its philosophy and application.

Important   A number of chapters in this book end with a table listing the principal advantages and disadvantages of the various processes previously described. It is essential that the contents of these sections be studied, as therein lie many examination question solutions!

Bibliography

Bicheno, J. 1991. *JIT cause and effect – a pocket guide*. Buckingham: Picsie Books.

Gould, P. 1992. *Concurrent engineering for survival, professional engineering*. London: Institution of Mechanical Engineers.

Kief, H. B. & T. F Waters 1992. *Computer numerical control*. Westerville, OH: Macmillan/McGraw-Hill.

Lambert, I. 1994. *Successful implementation of concurrent engineering products and processes*. New York: Van Nostrand Reinhold.

Wickham, T. 1993. Time is of the essence. *Sunday Times* (17 October).

World Bank 1994. *World development report*. New York: World Bank/Oxford University Press.

Chapter Two
Shapes made from molten metals – casting

After reading this chapter you should understand:
- (a) what is meant by casting
- (b) the basic requirements of any mould or die
- (c) the major differences between permanent and non-permanent moulds
- (d) why sand casting is the most commonly used general-purpose casting process
- (e) why different types of sand are used in sand moulding
- (f) typical casting defects and why they occur
- (g) what the shell moulding process is
- (h) why full mould casting is particularly attractive to development engineers
- (i) why investment casting is used for difficult to machine metals
- (j) why one die-casting method will not suffice for all non-ferrous metals
- (k) the special benefits of centrifugal casting
- (l) why over 90 per cent of all steel produced is now made by continuous casting
- (m) some of the most important basic rules of casting design.

2.1 The basic casting principle

One of the most popular methods of producing parts in metal is by casting. Casting is the process of forming objects by pouring liquid metal into a cavity having the same shape as the finished article (the mould), and then letting it solidify and cool. When removed from the mould, the casting produced should be an exact replica of the mould.

Exercise   Where else is this principle used? Do not confine your thinking to engineering.

2.2 Main requirements of any mould

There are many ways of making a casting, but all moulds must
- (a) have the desired shape and size but with allowances made for metal contraction during solidification and cooling (for steels this is typically 2 per cent, for cast iron 1 per cent and for aluminium 1.25 per cent)
- (b) be made from a material that is capable of withstanding the inrush of molten metal before solidification occurs (molten metal is just as heavy when liquid as it is when solid!)

(c) be designed to ensure efficient feeding of metal to all parts of the mould before it begins to solidify

(d) not cause a restriction to solidification and contraction of the metal after pouring

(e) allow easy escape of gases from the mould during pouring

(f) not cause undue difficulties in removing the casting after cooling.

2.3 Non-permanent and permanent moulds

Depending upon the number of identical castings required, their size and the metal being cast, the production engineer must decide whether to use a mould that can only be used once (non-permanent mould) or one that can be used repeatedly to make large numbers of castings (permanent mould). There are no rigid guidelines as to which type is preferable for any given situation, but in most cases the choice is normally obvious, and a study of the most commonly used methods of casting described in this chapter should make this clear. Generally, with the exception of continuous casting of steel, when large quantities of medium or small castings in non-ferrous metals are required, permanent mould casting is normally employed. Otherwise non-permanent moulds are used.

2.4 Non-permanent mould casting methods

Most non-permanent moulding processes involve similar basic steps, and as sand casting is the most common of these casting methods it is described below (§2.4.1) in considerable detail. It is strongly recommended that the reader carefully works through each step with the aid of the many diagrams provided (Figs 2.1 and 2.2), as a sound grasp of sand casting will make all subsequent casting processes described in this chapter and elsewhere much easier to understand.

2.4.1 Sand casting

As its name implies, this process uses sand as the material to make the mould cavity, but precisely how it is used in the construction of a mould is best explained by two examples. The first mould described is for the manufacture of an extremely simple component (component X), while the second example (component Y) introduces the reader to techniques involved in the production of slightly more complex shapes.

2.4.1.1 The production steps – component X (Fig. 2.1a)

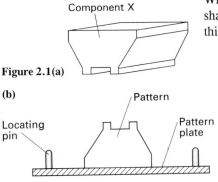

Component X

Figure 2.1(a)

(b)

Locating pin

Pattern

Pattern plate

While it is easy to say that the mould cavity must be made the same shape as the required part, for anything but the simplest of forms this is not always simple to achieve.

1. Make an accurate pattern (copy) of the required article in either wood or plastic. As stated in § 2.2, it must be slightly larger than full size to allow for contraction of the molten metal as it solidifies and cools to ambient temperature. Any areas that require machining after casting must also have sufficient metal available to allow for this (known as machining allowance). Small and medium-sized patterns, of the sort described in this example, are usually screwed to a flat wooden or metal surface known as a pattern plate (Fig. 2.1b).

Fig. 2.1(c)

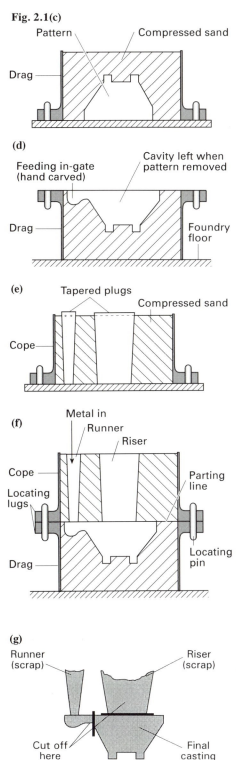

(c) Pattern / Compressed sand / Drag

(d) Feeding in-gate (hand carved) / Cavity left when pattern removed / Drag / Foundry floor

(e) Tapered plugs / Compressed sand / Cope

(f) Metal in / Runner / Riser / Cope / Parting line / Locating lugs / Drag / Locating pin

(g) Runner (scrap) / Riser (scrap) / Cut off here / Final casting

2. Place a square (or sometimes) circular steel surround on the pattern plate, enclosing the pattern (Fig. 2.1c). This surround is termed the drag or bottom box, although the latter is something of a misnomer as the box has neither a top nor a bottom.

3. Fill the space around the pattern with special moulding sand (§2.4.1.3), and ram it up as densely as possible – like compressing the sand in a child's bucket when making a sand castle (Fig. 2.1c).

4. Turn the complete assembly through 180° (Fig. 2.1d), and carefully remove the pattern, making sure not to damage the sand impression.

5. Using a hand trowel, carefully carve out a channel in the sand to provide a path along which the molten metal can flow into the mould cavity – this is called the feeding-in gate (Fig. 2.1d).

6. Using a steel surround identical to that used in step 2, place two tapered wooden plugs in position so that one will, in due course, align with the feeding-in gate in the drag and the other will align with the main mould cavity (Fig. 2.1e). Fill and ram up with sand. This assembly is called the cope or top box.

7. Carefully remove the two plugs from the cope, leaving two (tapered) holes in the compacted sand. These holes are termed the runner and riser.

8. Gently lower the cope onto the drag, making sure they are correctly aligned by placing locating pins through the holes in the locating lugs on each side of both cope and drag (Fig. 2.1f). (The joint/interface between the two halves of the mould is called the parting line.)

9. Pour the molten metal into the mould through the runner hole (Fig. 2.1f), and continue until it is seen coming up the riser hole. The pourer can then be sure that the mould is full. Leave to solidify and cool.

10. Transport the complete mould assembly to the knock-out area (an area specially prepared for removal of castings from their moulds, and which usually incorporates a system for feeding used sand back to a sand reclamation plant). Remove the casting from the mould (Fig. 2.1g). This is usually carried out with the aid of a mechanical vibration device.

11. Cut off the runner and riser (Fig. 2.1g), and remove any sand adhering to the surface of the casting by shot blasting or other mechanical means – a process known as fettling. The casting is then complete and ready for inspection and any heat treatment or machining that may be required.

Note The size of the riser in relation to the casting illustrated in Figure 2.1g is not exaggerated. The reason for its large volume relative to component size is to make certain that it does not solidify before the component does. The casting can then continue to draw molten metal from this reservoir of liquid metal as it shrinks and until it solidifies. This avoids porosity and sponginess problems (§2.4.1.4) and explains why yield rates (component volume to volume of metal cast) of only 50 per cent are so common in sand casting foundries.

Unfortunately most components are not quite such a simple shape as component X; most castings, for example, contain hole(s) of various configurations. They also have profiles that present problems in removing the pattern from the drag. An example (component Y) that embodies both of these problems is illustrated in Figure 2.2a, and would be made in the following way.

2.4.1.2 The production steps – component Y (Fig. 2.2a)

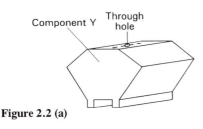

Figure 2.2 (a)

In a sand mould, all areas where there is no sand will fill with molten metal. Therefore if a hole in a casting is required, it is only necessary to put some sand (called a core) having the shape of the required hole at the location where the hole is needed. For example, if a circular hole is required a compressed sand core in the form of a cylinder would be used. (Small holes are usually more efficiently produced by drilling after casting rather than by coring.)

Exercise Try to get access to the cast iron cylinder head of a not so modern car – most cars these days have aluminium alloy die cast (§2.5.2) heads. Examine it closely and note just how complex the water passage holes are, how difficult it must have been to support the cores used to cast these holes and how difficult it is to ensure that all traces of core sand have been removed, so preventing sand from entering the cooling circuit when the engine is first run.

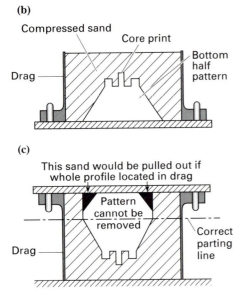

Because it is difficult to insert cores accurately and, equally important, to prevent them from being displaced when subjected to the forces associated with the inrush of molten metal at the time of pouring, cores are placed in specially recessed areas made in the sand of the main mould profile. These recesses are called core prints and are made by appropriate shaping of the pattern (Fig. 2.2b). Another problem with more complex castings is that it may not be possible to contain the entire shape solely in the drag because, as with component Y, their profile may be such that it would be impossible to remove a one-piece pattern from the mould after sand ramming (Fig. 2.2c). In such circumstances it is necessary to split the mould in a different place, and have part of the profile in the cope and the rest in the drag, i.e. relocate the parting line. The moulder cannot do this, of course, and it is the pattern-maker's decision where and how best to split the pattern. He then makes the two halves of the pattern accordingly.

Figure 2.2 (d)

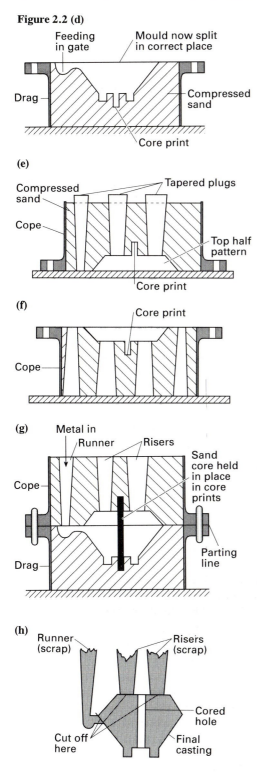

(e)

(f)

(g)

(h)

1. As with component X, make the drag by filling the area between its inner perimeter and the pattern with sand and then compacting it. Then turn the drag through 180° and carefully remove the pattern (Fig. 2.2d).
2. With a trowel hand carve a feeding-in gate (Fig, 2.2d).
3. Now make the cope. In contrast to component X, for component Y it must also contain a portion of the shape of the casting. Its construction is therefore slightly more complex than was the case in the previous example, and involves a similar procedure to that used when building up the drag. Place the cope around the top half of the two-part pattern, locate tapered plugs to provide runner and riser holes and finally fill the whole box with sand and ram it up (Fig. 2.2e). Then carefully remove the two plugs.
4. Next turn the cope through 180° and carefully ease the pattern out of the compacted sand (Fig. 2.2f), taking special care to ensure clean holes through from the riser holes to the cavity forming the upper part of the casting. Once again turn it through 180°.
5. Carefully place a previously made sand core in the core print in the drag, and then gently lower the cope onto the drag, making sure that the other end of the core fits correctly up into the core print hole in the cope. This ensures that the core is securely held at both ends (Fig. 2.2g). As in the previous example, precise location of both cope and drag is assured by the use of precision locating pins in the lugs on the outside of the two moulding box halves.
6. Pour the metal into the mould until it is seen to be full, and then let it solidify and cool. Figure 2.3b clearly shows the process of metal pouring.
7. Next remove the casting from its moulding boxes as previously described but, in contrast to component X, ensure that all traces of the sand core are removed during fettling, leaving a clean hole through the casting.
8. Finally, fettle the casting ready for inspection and any other operations called for on the component drawing (Fig. 2.2h).

(a)

(b)

Figure 2.3 (a) Two moulds ready for pouring and one left open to show core *in situ*. (b) Pouring molten metal into a sand mould. (Courtesy of William Cook plc, Sheffield.)

2.4.1.3 Types of sand used in sand casting

The main types of sand used in sand casting are green sand, CO_2 setting sand and core sand.

(a) *Green sand*. (The term "green" does not refer to its colour – it is actually black.) If one examines a mould ready for pouring (Fig. 2.1f or Fig. 2.2g), it is clear that the parts of the mould cavity that determine both casting shape and surface finish are the sand at the surface of the mould cavity and the outer surface of the core, i.e. the sand that actually comes into contact with the molten metal. Obviously the rest of the sand in the cope and drag is necessary, but only in a structurally supporting role to the sand at the cavity surface. It consists of silica sand, mixed with approximately 3 per cent coal dust (which gives it its characteristic black colour), 6 per cent clay and 3.5 per cent water to act as a strength-enhancing binder (British Cast Iron Research Association 1993).

In the past, to ensure the best surface finish possible, it was common practice to use new sand in the immediate area around the pattern and reclaimed (backing) sand as a filler to top up the moulding boxes. Owing to increasing cost-consciousness in the foundry industry and major improvements in efficiency of sand reclamation units, this practice is now rarely used, and the boxes are nowadays normally filled entirely with reconstituted sand of adequate uniformity and fineness.

(b) *CO_2 setting sand*. A more expensive alternative to green sand is CO_2 setting sand. This is a sodium silicate-based sand that chemically hardens when CO_2 gas is passed through it. It gives a much harder moulding surface than green sand, but because of its cost and use-only-once characteristics it tends to be reserved for costly alloy steel casting work. This is because molten steel is much less fluid than cast irons and non-ferrous metals, and so imposes larger forces on the sand during pouring. Cores can also be made from this type of sand.

(c) *Core sand*. In contrast to green sand, core sands are clay free, but they still need a binder. This is usually a resin that hardens with the application of heat. Thin, complex cores often require reinforcing wires to give them sufficient structural strength to withstand handling during mould assembly and to resist the forces imposed by the molten metal during pouring. Thus, core sand must be capable of binding well to such reinforcement.

Despite the differences between the various sands used in moulding, all sands must have a uniform consistency. They must flow freely as well as have sufficient strength when rammed up to retain their shape, particularly when subject to the forces imposed by the incoming liquid metal (*cohesivity*). Sand grains must not fuse together when they come into contact with liquid metal (*refractoriness*), as this would destroy the mould's *permeability* – the ability to let gases escape between the sand grains during pouring. Finally, moulds must crush in compression (*collapsibility*), so that as the metal solidifies and then cools the resulting contraction can continue unhindered by any part of the mould, including cores, which by that time are virtually redundant anyway.

2.4.1.4 Typical sand casting defects and their causes

Scabs
Excrescences on casting surfaces due to small portions of moulding sand breaking away during pouring.
Causes Poor ramming; inferior quality of sand used; incorrect in-gate placement.

Blowholes
Significant-sized, irregular-shaped holes running into and/or through the casting.
Causes Insufficient venting of the mould, possibly caused by too severe ramming; insufficient number of risers; mould not thoroughly dried before pouring, causing gassing; metal temperature too low during pouring.

Hard spots
Areas of the casting's surface have varying levels of hardness.

Causes Too rapid cooling in localized areas, usually in thin sections; incorrect use of chills (small pieces of steel built into selected points of the mould to speed up cooling in localized areas, and thus even out cooling across the whole casting).

Note Steel chills stay in the casting permanently unless they happen to be sited in areas that are subsequently machined. Their permanent existence in iron castings does not cause any weakness, although with steel castings any chills used must, of course, have at least as good metallurgical properties as that of the metal being cast.

Porosity and sponginess

Casting porosity, usually affecting heavy, thick sections.

Causes Incorrect metal composition such as too much phosphorus; too few risers or risers not large enough or incorrectly placed to allow them to feed the casting adequately with molten metal during solidification; slag inclusions in the metal. (Slag is the rubbish that floats on top of the molten metal before tapping the furnace. Careful pouring into the ladle usually avoids it getting into the mould.)

Cold shuts

Discontinuities within the casting – looks like stratified lava flow.

Causes Pouring process too slow; metal too cool when poured. Either of these causes can result in thin sections solidifying before the mould is totally filled and/or two molten metal streams meeting at too low a temperature for them to fuse together properly.

Displaced cores

Cored holes either the wrong shape or in the wrong place.

Causes Cores incorrectly sited, moved out of position during mould assembly or dislodged by the inrush of molten metal during pouring – often results in uneven wall thicknesses within the casting.

Fins

Thin slivers of metal (flashing) around the casting corresponding to the location of the mould parting line.

Causes Poorly fitting cope and drag, often the result of damaged mating surfaces; poorly fitting core prints. Can be an indication of poor foundry quality control.

Cracks

Fissures within the casting.

Causes Poor mould design in not allowing for contraction during cooling; mould too rigid and hard to allow for crumbling upon metal solidification; metal too hot when poured.

Centre-line shrinkage

Localized cracking of the casting along its centre-line – sometimes confused with porosity; usually worst in heavy sections.

Causes Outside surface cools and solidifies before the centre, so that when the centre does finally solidify it cannot be fed from any source of molten metal. This leaves a cavity in the casting's centre section. Incorrect or lack of use of chills in the affected areas can also cause centre-line shrinkage.

2.4.2 Shell moulding

As stated in section 2.4.1.3, the area of a sand mould that determines a casting's shape and surface finish is the sand surface that is exposed to the molten metal at the time of pouring. All the remaining sand in the mould exists purely to give strength and support to this critical interface. Clearly, the larger the casting, the greater the mass of metal poured into the mould and the stronger the mould must be to provide adequate support. Conversely, for small castings, up to approximately 20 kg, only a very modest amount of support is called for, and it is this fact that enabled shell moulding to be developed.

The process involves replacing the cope and drag of a normal sand mould with a pair of strong thin-walled shells which, when clamped together, produce an internal envelope corresponding to the shape of the required casting. The shells are still made of a fine sand but, unlike green sand, they are coated in a thermosetting resin similar to that used in most core sands.

Shell moulding involves the following steps:

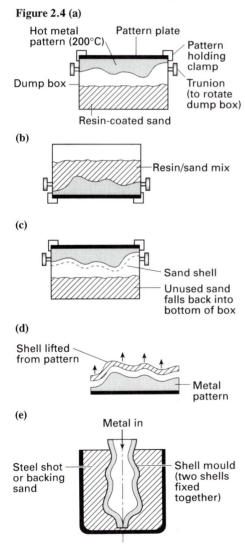

Figure 2.4 (a)

Hot metal pattern (200°C)
Pattern plate
Pattern holding clamp
Dump box
Trunion (to rotate dump box)
Resin-coated sand

(b)
Resin/sand mix

(c)
Sand shell
Unused sand falls back into bottom of box

(d)
Shell lifted from pattern
Metal pattern

(e)
Metal in
Steel shot or backing sand
Shell mould (two shells fixed together)

1. A metal pattern, attached to a pattern plate (usually cast iron), is heated in an oven to approximately 200°C.
2. Upon removal from the oven it is sprayed with a lubricant/releasing agent, and then attached to the top of a container (known as a dump box) containing resin-coated sand (Fig. 2.4a).
3. The dump box is rotated through 180°, causing the coated sand to fall onto the hot pattern (Fig. 2.4b).
4. When a partially cured layer (shell) of approximately 5 mm thick has formed around the pattern, the box is rotated to its original position, letting the unused sand fall back to the bottom of the box (Fig. 2.4c).
5. The pattern plate is removed from the dump box, and the shell (or biscuit as it is often called), still attached to the pattern, is placed in an oven for a further few minutes at 350–400°C to complete the resin curing process.
6. On removal from the oven the shell is carefully removed from the pattern, and is ready for use, or storage for use later (Fig. 2.4d).
7. Normally most components made via the shell moulding process require the use of two half-shells (like the cope and drag in sand casting). In such cases two shells are either glued (Fig. 2.5a) or wired together, and the shell mould is then ready for pouring (Fig. 2.5b). Where larger shell mouldings are involved, it is usual to provide additional support to the shell mould by enclosing it in an open-top container filled with steel shot or moulding sand (Fig. 2.4e).

When the molten metal comes into contact with the shells, its heat burns off the resin binder. Fortunately, this happens sufficiently slowly that the shells do not disintegrate before the metal has solidified enough to retain its shape during the cooling down process. This also has the advantage that the sand from used shells is easily removable from the surfaces of finished castings.

(a)

(b)

Figure 2.5   (a) A pair of shells, complete with a core, being assembled. (b) Shells glued together and ready for pouring. (Courtesy of William Cook plc, Sheffield.)

2.4.3 Full-mould casting

Two fundamental requirements of sand moulding are costly patterns in either wood, plastic or metal and the ability to withdraw patterns from the mould without damage to either mould cavity or patterns before metal can be poured.

The second of these requirements places a severe restriction or, at best, considerably complicates the production of castings that have complex shapes (Fig. 2.2c). However, by employing the full-mould (evaporative polystyrene pattern) casting process both of the above requirements are eliminated, and because of the speed with which a mould can be made it is particularly attractive to research and development departments whose work frequently takes second place in pattern shops to their day-to-day production work.

Despite the need for a new pattern for each component cast, full-mould casting is no longer restricted to small batches, as advances in high-speed polystyrene pattern production now make it commercially viable for large-quantity production of castings weighing up to 250 kg.

Full-mould casting, in its simplest form, involves the following steps:

Figure 2.6 (a)

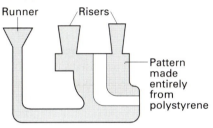

Pattern made entirely from polystyrene

Figure 2.6 (b)

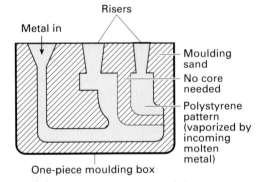

One-piece moulding box

1. A pattern of the required component is made from expanded polystyrene using simple hand tools. The various pieces needed to make up the complete pattern are then glued together, complete with runner, risers and feeding-in gate system, which are also made from polystyrene (Fig. 2.6a).
2. The completed polystyrene pattern is then placed in a one-piece moulding box, and surrounded by moulding sand (Fig. 2.6b).
3. Finally the molten metal is poured into the mould, instantly vaporizing all polystyrene in its path, and filling the cavity left by the vaporized pattern.
4. After solidification and cooling the casting is retrieved from the one-piece moulding box, fettled and inspected in the normal way.

2.4.4 Investment casting – lost wax casting

The advent of the gas turbine and the almost simultaneous development of missiles introduced new manufacturing problems, because of their need for extremely complex and accurate shapes made from difficult-to-cut materials. To meet these demands a modified form of a casting process, used by dentists to make gold fillings, was developed (British Standards Institution 1992). Despite having no connection with the world of finance, it is termed investment casting. It is also known as the lost wax process, the reason for this being evident after studying the following process details.

Investment casting involves the following steps:

1. The first stage in producing the required part is, as usual, to make a pattern. In this process the pattern is made from a low melting point wax, and because it has such a low melting point (90–100°C) it is easy to cast the pattern in an aluminium die.

Note   With very few exceptions, a mould made from metal is referred to as a die. If two or more complementary die parts are required to make a complete (metal) mould assembly, i.e. like the cope and drag of a sand mould, this is known as a die-set.

Figure 2.7 (a) Pattern

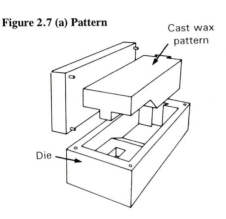

Cast wax pattern

Die

2. When the wax has solidified, the die is opened and the wax pattern is removed (Fig. 2.7a). As it is normal and more economic to make several components in one mould, a number of wax patterns are required, but as the method of pattern casting is so simple this is not a problem.

3. When sufficient wax patterns have been made they are then assembled (gated) onto a central runner or sprue (stem), to form a tree-like structure (Fig. 2.7b).

4. This multipattern wax assembly is next submerged in a vat containing a fine ceramic-based slurry known as investment, hence the name of this casting process (Fig. 2.7c). It should be noted that the fineness of this slurry determines the surface finish of the finished components. It is then coated (stuccoed) with a coarse refractory powder (Fig. 2.7d) until a coating thickness of 5–10 mm has been built up over the whole wax assembly.

5. When the investment has set and dried it is heated sufficiently for the wax tree to melt and run out (dewaxing), leaving cavities of the exact shape of the wax pattern in the investment (Fig. 2.7e). Virtually 100 per cent wax recovery is achieved and can be continually re-used.

(b) Assembly

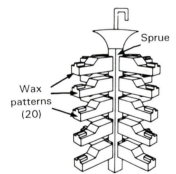

Sprue

Wax patterns (20)

(c) Investing

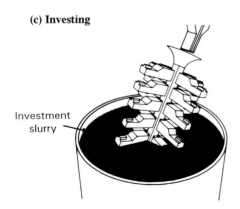

Investment slurry

(d) Stuccoing

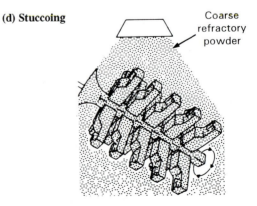

Coarse refractory powder

Exercise   Can you see a strong similarity here between investment casting and full-mould casting?
Hint   Consider what happens to the pattern when the mould cavity is created.

(e) Dewaxing

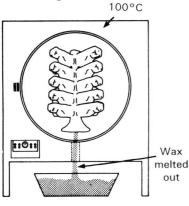

100°C

Wax melted out

6. The investment mould is next placed in an oven and heated to approximately 1000°C to vaporize any last vestiges of wax that may be lurking within the investment mould cavities, and to fully harden off the investment (Fig. 2.7f).
7. The mould is removed from the oven and, while still hot, the molten metal is poured into it (Fig. 2.7g).
8. When solidified, the cast tree of components is retrieved by breaking away (knocking out) the brittle investment material (Fig. 2.7h).
9. Finally, each component is carefully cut from the tree, fettled and inspected before any subsequent machining processes that may be required.

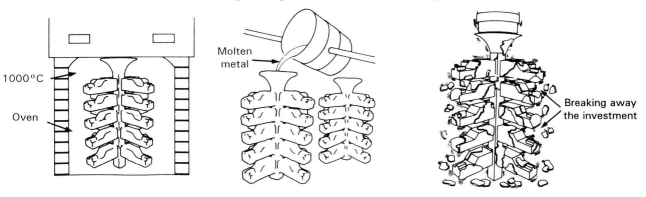

(f) Firing

1000°C

Oven

(g) Casting

Molten metal

(h) Knock-out

Breaking away the investment

2.5 Permanent mould casting methods

All the moulding processes described so far have one thing in common – they require a new mould for every casting. Such a wasteful situation is acceptable for limited numbers of identical components, but when large-quantity production is required, great savings are possible by investing in moulds that can be used over and over again, i.e. permanent moulds.

Unfortunately the biggest problem is finding a mould material that will withstand the melting points of ferrous metals – about 1200°C for cast irons and 1550°C for steels. For cast iron and most non-ferrous metals, steel dies are used, although when pouring cast iron frequent redressing (smoothing) of the die-set profile is usually necessary. Steel is occasionally cast in graphite moulds, but their production is so difficult when complex shapes are involved that steel die casting is rare compared with non-ferrous die casting, in which the dies are made from alloy

Figure 2.8 Turbocharger rotor and the wax pattern used to make it. (Courtesy of Trucast Ltd, Ryde, Isle of Wight.)

(a)

(b)

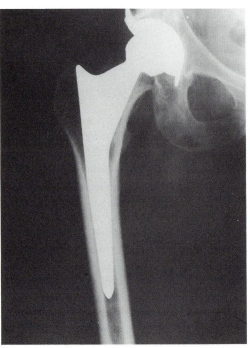

Figure 2.9 (a) Hip joint and sockets cast by lost wax process. (b) Radiograph of a hip joint in position. (Courtesy of Trucast Ltd, Ryde, Isle of Wight.)

steel. Therefore, if quantities of precision parts in high melting point metals are required, investment casting (§2.4.4) is invariably the best option.

Another feature of permanent moulds is that, by definition, they are not destroyed while getting the casting out. Therefore all permanent moulds must be capable of being opened and closed repeatedly for casting removal. This complicates mould design considerably, particularly when cores are necessary, and this is why most components made by permanent moulding methods tend to require fairly straightforward coring systems, although there are exceptions.

Steel inserts are frequently used in non-ferrous castings to enhance strength in localized areas. They are similar to those used in plastics moulding, and are illustrated in Chapter 4 (Fig. 4.8).

2.5.1 Gravity die casting

This is the simplest form of permanent mould (die) casting and, in effect, involves the replacement of a sand mould with a fine-grain cast iron die-set. However, because the die is permanent, provision must be made for the die to be easily opened after every pour to release the casting produced. Also, the locking force needed to hold the parts of the die together must be sufficient to ensure that no molten metal can seep out through the die joint line(s), as this would leave "flashing", which would then have to be removed by hand. In contrast to high-pressure die casting (§2.5.3), as the molten metal only flows into the die cavity under gravitational force (Fig. 2.10), this die-set closure force is modest.

Great care must be taken in venting the die cavity if gaseous porosity is to be avoided in the castings produced. This is a much greater problem with metal dies than with sand moulds as the latter have a degree of inherent permeability through which the gases present at the time of metal pouring can escape. Remember, steel dies have zero permeability!

The batch sizes typical of this process suit manufacture of such items as high-strength magnesium alloy car and motor cycle racing wheels, but large-volume aluminium alloy wheels fitted to certain production vehicles are best produced by low-pressure die casting (§2.5.2).

Sand cores can also be used in gravity die casting, making complex internal forms possible without a disproportionate increase in die-set costs.

2.5.2 Low-pressure die casting

This is a slightly more complex die-casting method, but has the advantage that runners are not required, thus offering a greater casting yield for a given volume of metal cast.

The process involves replacing gravity-fed metal into the die with a pneumatic force derived from a low-pressure compressed air source – typically 0.5–1 bar. The air supply is applied to the surface of the molten metal, and this forces it up a ceramic-lined feeder tube into the die cavity (Fig. 2.11). Not having to remove runners or risers reduces fettling costs, as well as minimizing machining requirements. Casting quality is also much more consistent than is achievable with gravity die casting. Die life is normally at least 50 000 castings (Palmer et al. 1981).

Aluminium alloy wheels, multivalve engine cylinder heads and similar complex-shaped parts are typical of the products made by this process.

Figure 2.10 Metal pouring into a gravity-fed die-set. (Courtesy of Alumasc Ltd, Burton Latimer, Northants.)

Figure 2.11 Principle of low-pressure die casting. (Courtesy of Alumasc Ltd, Burton Latimer, Northants.)

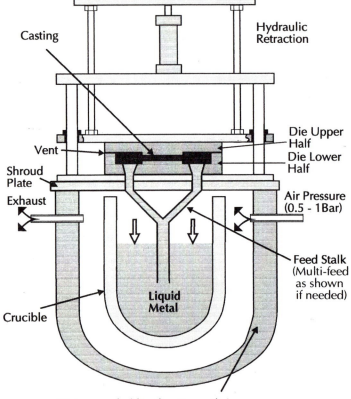

2.5.3 High-pressure die casting

If solidification/cooling time could be significantly shortened a major increase in die-casting productivity would result. This is actually possible, and is achieved by using a die-set fitted with efficient water cooling.

Unfortunately, there is metallurgical incompatibility between steel and molten aluminium and its alloys. This results in the need for two versions of high-pressure die casting – the hot chamber process (§2.5.3.1) and the cold chamber process (§2.5.3.2).

2.5.3.1 Hot chamber die casting – not suitable for aluminium and its alloys

High-pressure die casting relies upon the action of a piston to deliver molten metal into the die cavity from a storage reservoir (the hot chamber from which this process derives its name).

This piston has a four-stage operating cycle:
1. closing off the metal feed from the hot chamber reservoir
2. moving the liquid metal charge up a feeder tube to the die's feeding-in gate
3. injecting the molten metal into the die cavity as quickly as possible
4. intensifying the injection pressure on the injected metal to compact it and eliminate any risk of porosity forming in the casting during solidification (Fig. 2.12).

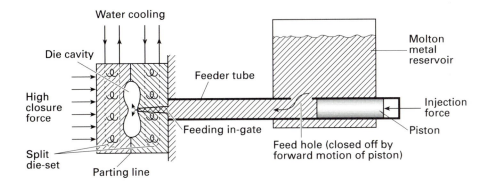

Figure 2.12   Hot chamber die casting principle.

On completion of this cycle the piston retracts, the die is automatically opened and the casting ejected. The die-set then recloses and the feeder tube is again filled with the correct volume of charge ready to make the next casting.

As the die-set is efficiently water cooled, the metal solidifies extremely quickly and the casting can therefore be ejected from the die with equal speed. Small components can be made so rapidly that production rates measured in seconds are readily achieved. For such short cycle times to be sustained, casting removal must be automated, which is also convenient as, for safety reasons, all high-pressure die-casting machines are heavily shielded to make human access difficult.

Gravity and low-pressure die casting both require an adequate closing pressure between the parts of the die-set to avoid metal leakage along parting line(s), but this is even more critical with high-pressure die casting owing to the extremely high injection forces used in this process – up to 1500 tonnes.

Exercise Carefully examine a modern car or motor cycle engine. There are a number of major components that have been made by die casting. When you have located them, see if you find any evidence of where the die-set parting line(s) were located. If you cannot, both the designer and moulder have done their jobs well. Carry out a similar inspection of some common domestic appliances, such as a washing machine, and toy models.

Even greater integrity of die castings can be achieved by carrying out high-pressure die casting under vacuum. This virtually eliminates any risk of casting porosity and enhances metallurgical purity.

Hot chamber vacuum casting is not difficult to engineer as the whole process is highly automated and enclosed to prevent any possibility of human contact with high-pressure molten metal.

2.5.3.2 Cold chamber die casting – suitable for aluminium and its alloys

Unfortunately, when molten aluminium or its alloys come into contact with cold ferrous (iron and steel) surfaces they tend to absorb iron, causing both pitting of the steel die surfaces and contamination of the molten metal. A solution to this problem is to minimize the time that any ferric-based parts of the moulding machine and die-set are in contact with the molten metal. This can be achieved by modifying the hot chamber process and moulding machine in two ways.

First, the molten metal is no longer held in a heated reservoir adjacent to the die, but is kept away from the moulding machine. Each charge is individually ladled into the feeder tube when required. Secondly, the feeder tube and the end of the feeder piston, both of which come into unavoidable contact with the molten metal, are lined with a protective high melting point non-ferrous nickel-based super-alloy coating. All other aspects of the cold chamber injection moulding process (Fig. 2.13) are the same as in the hot chamber process.

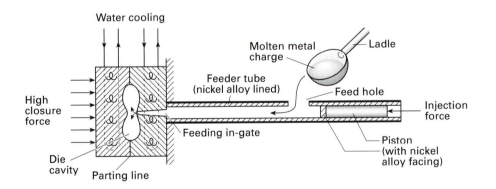

Figure 2.13 Cold chamber die casting principle.

Clearly, the mechanical ladling of each molten charge significantly slows down the rate of production, but at least it does enable aluminium and its alloys to be die cast satisfactorily. As expected, this increased cycle time is reflected in component production costs, but if only high-value components that demand aluminium are involved this extra labour cost is of much less significance than would be the case with high-volume, low-cost zinc-based parts typical of hot chamber production.

2.5.4 Centrifugal casting

The casting processes so far described in this chapter use various driving forces to propel molten metal into a mould or die cavity. They include gravitational force (used in all non-permanent mould methods and in gravity die casting), compressed air pressure (used in low-pressure die casting) and mechanical ram force (used in high-pressure die casting). Centrifugal casting, as its name implies, employs centrifugal force.

Like most casting methods, the principle involved is simple. Molten metal is poured into a rotating mould (Fig. 2.14); as a result of centrifugal force, it is then flung towards the outer surface of the mould. The thickness of the casting is controlled purely by the volume of metal fed into the mould, and not by the use of a core. Runners and risers are also unnecessary.

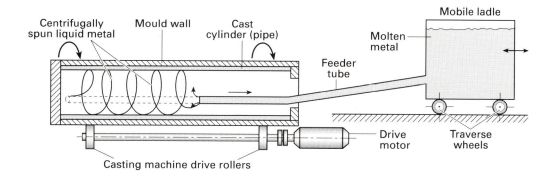

Figure 2.14 Horizontal centrifugal casting.

Rotational speeds vary from 300 to 3000 rev/min, depending upon the internal radius of the mould, but a centrifugal force of approximately $75\,g$ is normal.

While this is clearly a process ideally suited to the production of circular shapes such as thick-wall piping, variations to this principle do permit other shapes to be centrifugally cast (Niebel et al. 1989). The axis of rotation is usually horizontal for objects with large length/diameter ratio (e.g. pipes) and vertical for items with small length/diameter ratio (e.g. railway rolling stock wheels and brake drums).

Most metals can be cast, although when iron and steel are used the mould is lined in sand or other refractory material. This is not necessary when casting non-ferrous metals due to their lower melting points.

The benefits and limitations of this process are listed in §2.8 at the end of this chapter.

2.5.5 Continuous casting

All the casting processes so far described are options open to the production engineer. However, there is one process that is not suitable for component manufacture, and this is continuous casting. Because of its unprecedented increase in popularity since the early 1980s (British Steel now makes over 90 per cent of all its steel by this process; British Steel 1992), this chapter would be incomplete without the inclusion of a brief description of the basic principles involved.

The reader will by now have been conditioned to think of moulds and dies as containers into which molten metal is poured or injected and allowed to solidify before removal (i.e. discrete component production). Indeed, this is the basis of all the casting processes so far described, but continuous casting is very different, and demands a change of thinking.

Like all casting processes, continuous casting begins with molten metal as the source material but, unlike other moulding methods, it is not poured directly into the mould. Instead it first passes to a reservoir/settling tank called a tundish, which acts as a buffer store (reservoir) for the liquid metal (Fig. 2.15). This is necessary because, although this process requires a continuous stream of molten metal, continuous melting, for various reasons, is impractical. Also, with a residence time of about 10 minutes in the tundish, this allows any slag present to float to the surface and not be poured into the mould.

Out of the bottom of the tundish flows a steady stream of molten metal into the mould called a strand. This mould in very different to any so far described, and consists of a water-cooled copper alloy box having the same cross-section as the strand to be cast (Reynolds 1993), but it has no bottom. The liquid metal therefore flows into the mould from above and, owing to the very rapid and efficient cooling action around the copper mould walls, it can be drawn out of the bottom of the mould as a continuous length of metal whose outside surface has solidified just sufficiently to retain its basic shape. The centre of the strand is still molten at this stage (like a

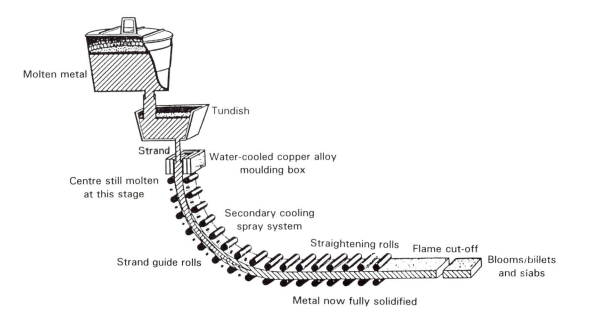

Figure 2.15   Essential elements of the continuous casting process. (Courtesy of British Steel plc.)

soft-centred lozenge), and therefore secondary cooling in the form of water sprays is essential to solidify the strand over its whole cross- section.

If the primary cooling around the mould is inadequate, or if the time that the metal takes to pass through the mould is too short, the outer skin will not have solidified sufficiently to retain the still liquid centre. Molten metal will then break out, spill and solidify over adjacent parts of the casting machine, causing a serious and expensive breakdown.

Finally, the strand passes through guide and straightening rolls to an oxy-gas flame cutter, which cuts it into desired lengths (Fig. 2.15).

Increasingly, other operations are added to this casting process, such as hot rolling (§3.2.1). This is not surprising as, with the strand still being at a sufficiently high temperature to permit hot working (§3.1.1), this avoids the alternative expensive option of letting discrete lengths cool down, and then having to reheat them again when they are required for rolling later.

Non-ferrous metals such as aluminium and copper, are also continuously cast, but by far the greatest tonnage poured involves the production of steels.

2.6 Transporting and pouring the molten metal

The melting furnace and the ladle used to transport liquid metal to individual moulds must have sufficient capacity to ensure that the complete casting can be poured in one operation. Furthermore, the metal must be at a sufficiently high temperature when it is removed from the furnace to ensure that solidification will not commence before pouring is complete (typically 1600–1700 °C for steels and 1250–1300 °C for irons).

Handling molten metals is obviously an extremely dangerous activity and is only undertaken by people experienced in such work, and who are equipped with suitable protective clothing. One of the greatest dangers associated with non-permanent mould filling is that of wet moulds. If there is an excess of moisture in the mould cavity, when it comes into contact with the molten metal it is immediately converted into steam, and can cause droplets of molten metal to be blown back out of the mould, causing great danger to the pourers and other foundry personnel in the immediate vicinity.

2.7 Some important casting design criteria

Although it is the job of the pattern-maker/die-maker to decide how best to cast a given component, and of the moulder to decide where best to place runners, risers, etc., the engineering designer can still make a major contribution towards ensuring that good-quality, economical castings are produced by following a few basic design rules. Some of the more important of these are given below (Steel Castings Research and Trade Association 1968; Council of Ironfoundry Associations 1970).

    (a) *Never* use sharp corners – always use the largest radii possible, but not to the extent that you create "heavy section" problems (b).

Figure 2.16 (a)

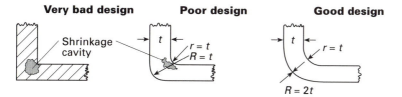

(b) Avoid "heavy sections" in localized areas, either by coring or by minor design modifications such as profile contouring.

Figure 2.16 (b)

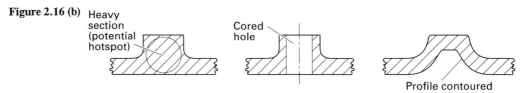

(c) As far as possible a casting should avoid any changes in section thickness, but if this is unavoidable never design rapid changes of section.

Figure 2.16 (c)

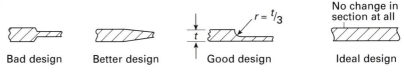

(d) Contraction allowances are necessary on all patterns, but this is always catered for by the pattern-maker working with a special contraction rule. Machining and distortion allowances must also be catered for, the amount depending upon casting size and configuration. As with contraction allowances, these are normally taken care of by the pattern-maker. Remember castings usually distort somewhat during cooling.

(e) Vertical sides should always be slightly tapered (i.e. have a draft angle) to simplify pattern removal from the mould – typically 1–3°.

Exercise Make a sand castle with a normal drinking cup with tapered sides. Now make one using a parallel-sided drinking beaker – a steady hand will be required! Can you see why cylindrical shapes do not require draft angles when cast horizontally?

(f) There are minimum practical limits to casting thickness. These vary with the casting process and the size of the component involved, e.g. pressure die castings can easily achieve 1 mm, but even moderate-sized sand castings have difficulty in achieving 3 mm consistently.

(g) Avoid re-entrant shapes whenever possible as they complicate mould/die design unnecessarily.

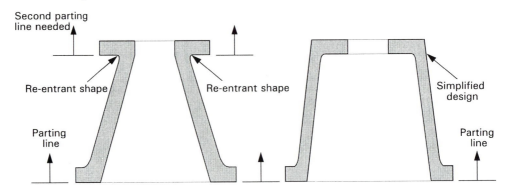

Figure 2.16 (d) Avoid re-entrant shapes.

(h) If a component's shape and size is such that a casting is clearly called for, do not overdesign it by specifying an unnecessarily expensive high-strength metal. Unless quantities or design parameters dictate, use the least expensive metal and the cheapest casting method that will meet the required specification. If cast iron is strong enough, and sufficient numbers are required to justify the cost of patterns, use it – there are numerous grades available with ultimate tensile strengths ranging from 150 to 400 MN/m^2. Cast iron is the least expensive and readily castable metal available, although it should not be used where shock loads are possible or in extremely cold environments as it becomes brittle at subzero temperatures. If cast iron is not strong enough, or if its other metallurgical properties are not suitable, then steels should be considered, although steel castings are much more expensive and difficult to make than cast iron ones. Non-ferrous metals such as aluminium, magnesium, copper and zinc are all castable, but expensive unless large quantities are required. Therefore, unless specific properties of a particular non-ferrous metal are essential and/or the quantities to be produced justify a high-volume casting process such as hot chamber die casting, they are best avoided.

The number one golden rule of casting design
Always discuss your finished casting drawings with the foundry before issuing them.

2.8 Summary of the principal characteristics, advantages and disadvantages of casting processes

Process	Characteristics	Advantages	Disadvantages
Non-permanent mould casting			
Sand casting	Moulds made from special moulding sands contained within steel containment boxes	One-piece products of complex shape can be made at moderate cost, and in large quantities, so casting is often a better alternative to fabricated subassemblies A flexible process that can cope equally well with either large, heavy castings or small parts made singly or in clusters Castings can be subjected to a range of heat treatments for both stress relief and grain refinement. Surface hardness can also be varied within certain limits There is little material wastage as feeders and risers can be remelted and used again It is easy for the designer to place metal where it is needed	Each mould and core can only be used once, so time spent making them is lost after each cast Sand casting is not as accurate as certain other casting processes, either dimensionally or in terms of surface finish There is an economic break-even point for certain shapes (usually relatively simple ones), below which fabrication or machining from the solid is cheaper than making patterns, cores, moulds, etc. Unpleasant, dirty, smelly working environment Large floor space needed to accommodate moulds in the course of assembly and those ready for pouring

Process	Characteristics	Advantages	Disadvantages
Shell moulding	A sand casting process that uses thin, resin-coated sand shells instead of moulding boxes to produce the required casting envelope	The shell mould is more compact than its sand casting equivalent Only semiskilled labour is required, and the process lends itself to significant automation – one shell per minute is not uncommon Shells can be stored for extended periods before use – green sand moulds must be cast within hours of completion Storage space required is low as shells can be stacked on shelves like domestic crockery Repeatability and dimensional accuracy is high (0.1 mm) compared with sand casting, in part because of the fineness of the sand used in shell manufacture. Surface finish is also good for the same reason Low scrap rates compared with sand casting No need for heavy moulding boxes In-gate, runner and riser systems can be incorporated within the pattern design, making for more rapid shell mould assembly No wasted resin-coated sand; what is not used this time is used for making subsequent shells	Shell moulding is limited to small to medium-sized components, partly because of the maximum oven size readily available The cost of the metal patterns is high, so frequent batches of at least 100 are necessary to justify the cost. Also, any design modifications are much more difficult with metal patterns than with wooden ones The resin-coated sand is expensive (up to four times that of green sand), and cannot be used again other than as the backing sand to support large shells when cast in one-piece flasks (Fig. 2.4e)
Full-mould casting	A sand casting process in which the mould cavity is created by evaporating a polystyrene pattern at the time of metal pouring	Patterns are much lighter than wood or metal, so are easy to handle One-off patterns are easy and quick to make using simple hand tools and/or light machine tools such as a band-saw Separate cores are rarely required since the pattern is not removed from the mould Two (or more) part moulds are unnecessary, thus eliminating parting lines	The pattern is lost as soon as the metal is poured; a new mould *and pattern* are therefore required for each casting produced Polystyrene patterns are fragile and easily damaged in the foundry Casting surface finish is controlled by the fineness of sand used and the smoothness of the pattern's surface. It is usually similar to that achieved in normal sand casting

Process	Characteristics	Advantages	Disadvantages
Full mould casting (contd)		Polystyrene is a cheap material compared with metal or seasoned wood	
Investment casting	A precision casting process in which temporary patterns are made in low melting point wax. This wax is coated in ceramic which, when hardened and the wax melted out, becomes the final mould into which the molten metal is poured	Because the pattern is produced as a one-piece wax structure, no pattern removal is necessary (it is melted out), nor is there a need for separate cores or loose pattern pieces. Hence the most complex shapes can be readily produced Accuracy and surface finish are so good that in many cases subsequent machining is unnecessary - particularly useful when dealing with tough, difficult to machine metals such as the Nimonics used in gas turbine blade manufacture Multiple components can be produced in one cast The pouring operation is readily adaptable to vacuum casting, further enhancing casting quality and consistency	A complex process overall, involving many steps Limited to relatively small components such as turbine blades, replacement hip joints, etc. weighing up to approximately 15 kg Highly skilled labour is required, making the process even more expensive Mould materials are costly, and while the wax is re-usable the investment is not Manufacture of the aluminium die used to make the wax patterns can be expensive if the component has a complex shape

Permanent mould casting

Process	Characteristics	Advantages	Disadvantages
Die casting	A group of casting processes that employ re-usable metal (or occasionally graphite) dies. Mainly used for non-ferrous metals, which are either gravity poured or forced into the die cavity under pressure	The hot chamber process lends itself to rapid, economic, mass production using semiskilled labour Die casting is capable of producing castings to remarkably close tolerances (0.025 mm), and frequently this eliminates the need for subsequent machining Very thin sections can be produced (less than 1 mm), owing to the high injection pressures used, guaranteeing complete filling of the die cavity. Cored holes are both very accurately positioned and of precise dimensions as a	Mainly non-ferrous metals with melting points of no more than about 900°C can be die cast if heavy die wear is to be avoided The cost of *die sinking* (reproducing a component's shape in a solid block of metal) is very high because of the skilled labour required, although this is now less of a problem thanks to CNC machine tools The cost of the alloy steels used for die making is high Significant die design changes are virtually impossible Owing to high tooling

Process	Characteristics	Advantages	Disadvantages
Die casting (contd)		result of the steel construction of the die-set Exceedingly smooth surface finish is produced as a result of the highly polished surfaces of the die. This often eliminates the need for any surface finish treatments Injection moulding can readily accommodate the inclusion of inserts of other metals (e.g. steel studs), cast integral within the die casting The hot chamber process is easy to perform under vacuum, so minimizing the risk of casting porosity	costs, die casting can only be economically justified for mass production, e.g. runs of at least 20 000 If production is reliant upon die castings in such large quantities, the risk of die failure cannot be ignored. This usually means that two die-sets per component are held, one as a stand-by, thus making die costs even higher Parts that can be die cast are limited in both physical size and mass – typically up to 30 kg in light alloy Aluminium and its alloys can only be cast via the cold chamber process, so this biases the use of these materials towards relatively high-value components While casting porosity is a potential problem with all die casting methods, castings made via the cold chamber process are more prone to porosity than other die castings as a result of entrainment of gases when ladling the molten metal into the die. Vacuum casting reduces this problem, but is more difficult to achieve than with the highly automated hot chamber process
Centrifugal casting	A casting process involving molten metal being thrown to the outer surface of a rotating mould (usually of circular configuration), by centrifugal force	Special metals and unusual pipe sizes not normally available from rolling mills can be easily cast Like low-pressure die casting, no runners or risers are necessary Because slag, gases etc., inherent to some degree in all molten metal, are lighter than the liquid metal, they are centrifuged to the inner surface of the casting, whence they can then be removed if	Expensive heavy-duty equipment is needed to spin molten metal *safely*, so batch sizes must be large enough to justify the expense involved For thick-walled items solidification occurs simultaneously from both the inside and outside surfaces. This can lead to shrinkage defects midway between these two surfaces The process is limited

Process	Characteristics	Advantages	Disadvantages
Centrifugal casting		necessary by a simple machining operation With sufficient centrifugal force metal density can be significantly increased with resulting improvements in its physical properties No central core is necessary	mainly to cylindrical castings
Continuous casting	A casting process which produces a continuous supply of metal of constant cross-section from a bottomless copper mould; increasingly the process which precedes hot rolling operations	Highly consistent dimensional quality and metallurgical properties produced Little waste and slag inclusions are all but eliminated With a simple adjustable mould design, section changes are easy and quick to achieve, usually without even stopping the casting process Post-cast rolling is frequently tagged on to the end of the casting machine, so offering a major energy saving compared with traditional ingot casting and rolling Increased production rates, current steel slab output velocity being of the order of 2 m/min. Double-strand machines are also now used	Because cooling occurs from the outside inwards a shrinkage cavity is sometimes formed. This must be eliminated during subsequent hot rolling operations The moulding equipment is complex and therefore expensive Metal break-outs are costly and can cause considerable damage to the moulding machinery Casting thicknesses below 50 mm are currently difficult to produce consistently

Bibliography

British Cast Iron Research Association 1993. *Preparation of green sands*. Alvechurch: BCIRA.

British Standards Institution 1992. *Investment castings in metal – BS 3146*. London: British Standards Institution.

British Steel 1992. *Technology and competitiveness – steelmaking and casting*. Teesside: Teeside Laboratories.

Council of Ironfoundry Associations 1970. *A practical guide to the design of grey iron castings for engineering purposes*. London: CIA.

Niebel, B. W., A. B Draper & R. A. Wysk 1989. *Modern manufacturing process engineering*. New York: McGraw-Hill.

Palmer, J., P. Howard & M. Strowbridge 1981. Low-pressure diecasting. *Engineering*, 221, 1–8.

Reynolds, T. 1993. *Continuous casting machine design*. IGDS course in metals technology management – continuous casting module. Sheffield University, December 1993.

Steel Castings Research and Trade Association 1968. *The design and properties of steel castings*. Sheffield: BCRTA.

Case study

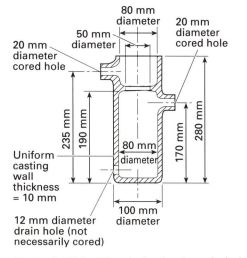

Figure 2.17(a) Filter body showing principal casting dimensions only.

A company has a requirement for 25 liquid filters per month, the principal component of which is its cast iron body (Fig. 2.17a). Suggest how this casting would be most economically produced, assuming that each month's demand is produced in a single batch.

Show where the pattern-maker is likely to locate the mould system parting lines.

Solution

As the material required is cast iron and the batch size is moderate, the most economical casting method is sand casting. Sand casting is invariably the cheapest method for small batch production and, as there are many foundries offering this process, prices are always competitive.

Wooden patterns and core boxes are adequate for such low annual production requirements, and are less expensive to make than the types of pattern required for most other casting options.

Mould configuration options

One's initial reaction might be to cast the component in the vertical position (Fig. 2.17b), as any slag, loose sand, etc. should then float to the top of the casting (face X), and be removed when this face is finish machined. Unfortunately, owing to the part's shape, if it is cast vertically, the mould will require three parting lines (Fig. 2.17b), i.e. cope, drag and two intermediate boxes. This dramatically increases pattern, core and mould assembly costs.

If the filter body is cast on its side (Fig. 2.17c), only one parting line is necessary, and this can be catered for by a basic cope and drag assembly. However, there is potentially a serious

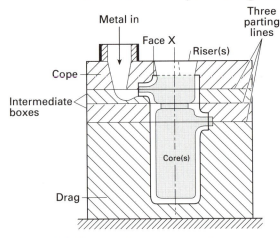

Figure 2.17(b) Filter body cast vertically – three parting lines.

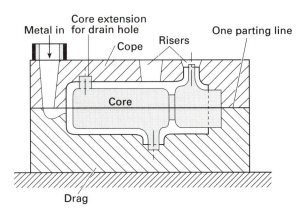

Figure 2.17(c) Filter body cast horizontally – one parting line.

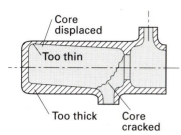

Figure 2.17(d) Core broken and displaced.

core structural strength problem when casting horizontally, because of its extended length (280 mm) in relation to its diameter (80 mm). While small metal core support devices called chaplets and core wire reinforcement (§2.4.1.3) would certainly be used, there is still a significant risk of the core cracking when its long unsupported length is subjected to the inrush of molten metal (Fig. 2.17d).

The result of this is likely to be that the core will be displaced (see §2.4.1.4), and this will cause the final casting to have a dangerously thin wall on one side and be excessively thick diametrically opposite (Fig. 2.17d). Fortunately, this filter design calls for a drain hole (Fig. 2.17a) located in just the area where the core is in most need of support. Therefore, instead of drilling the required 12 mm diameter hole after casting, if it is cored, the extended part of the core needed to form this hole can also act as a structural support for the main core at its furthest extremity (Fig. 2.17c).

While this makes mould assembly slightly more complex, it is a small price to pay to prevent core displacement or fracture. If a 12 mm diameter core was considered by the foundry to be too small to give adequate support, enlarging the drain hole size is unlikely to compromise the design of the filter. Indeed, a designer who had considered the full implications of casting this filter body when it was first designed would have realized that it should be cast horizontally, and would have incorporated an adequately large cored drain hole to aid core support. This is a good example of what is meant by "design for manufacture" (§1.2.1)!

Questions

2.1 What are the fundamental requirements of any process used to make metal castings?

2.2 Briefly discuss the major advantages of casting compared with other methods of basic shape production.

2.3 What properties are required of a good moulding sand? How does the sand differ from that used in core production?

2.4 Discuss the behaviour of metal cast into non-permanent moulds, making particular reference to casting hardness and grain structure.

2.5 Describe some common casting faults, and suggest steps that may be taken to minimize the risk of such faults occurring.

2.6 Compare the method of mould construction and the advantages, both technical and economic, of conventional green sand moulding with any alternative method of expendable mould construction.

2.7 Discuss methods employed to speed up the moulding cycle in the sand casting process.

2.8 By far the largest tonnage of ferrous castings produced annually is made by the sand casting process. Why is this? Discuss whether you consider that this situation is likely to continue in the foreseeable future, in view of the radical changes that have taken, and still are taking, place in the foundry industry.

2.9 What advantages have CO_2 and shell moulding techniques over conventional sand casting?

2.10 What is meant by full-mould casting? Why is it particularly attractive to development departments?

2.11 What are the main limitations of the investment casting process compared with alternative methods of casting?

2.12 The investment casting process is capable of producing components that would otherwise be difficult to manufacture. Give an example of such a component and explain the reason(s) for your choice.

2.13 What are the main similarities between the full-mould and investment casting processes?

2.14 Discuss the relative advantages and disadvantages of pressure and gravity die casting, stating which metals are most commonly used in each process. Why does vacuum die casting normally produce castings superior to those poured at atmospheric pressure?

2.15 Suggest the most suitable casting process to produce the following components. In each case give reasons for your choice:
  (a) small batches of cast iron engine cylinder liners
  (b) large quantities of zinc-based carburettor bodies
  (c) car engine connecting rods
  (d) gas turbine compressor blades
  (e) aluminium pistons for diesel engines
  (f) an aluminium-bronze ship's propeller.

2.16 State briefly advantages of the centrifugal casting process over normal gravity feed casting methods, and indicate the most significant limiting factor in its use.

2.17 In centrifugal casting, too low a speed will permit slipping or raining of the liquid metal, and too high a speed will result in hot tears in the outer periphery of the casting. A force on the outer radius of a sand-lined centrifugal mould (for casting ferrous metals) of about 750 N is found by experience to work well, and approximately 600 N is sufficient for unlined moulds (for casting non-ferrous metals) because of their greater chilling power.
  (a) Derive a relationship to express the centrifugal force existing in a casting shaped like an annular ring of outside diameter D, bore d, length L. Rotational speed is n rev/min.

$$\text{Centrifugal force} = (M \times v^2) \div R$$

  where M is the mass of the rotating metal, v is the rotational velocity and R is the radius of gyration.
  (b) For a metal ring of 150 mm outside diameter, 130 mm bore, and 100 mm long, calculate the mould speeds necessary to achieve the previously quoted forces when the ring is made from (i) low-carbon steel and (ii) aluminium. Take the density of low-carbon steel as 7850 kg/m^3, and 2560 kg/m^3 for aluminium.

2.18 Why do you think that over 90 per cent of all steel production in Europe is now made by the continuous casting method rather than by the traditional ingot casting process? Mention at least five advantages to be gained from using continuous casting.

2.19 Rolling equipment is increasingly being added downstream of continuous casting machines. Why is this?

2.20 Explain at least four essential design rules that must be adhered to when designing metal castings if problems, and possible high scrap rates, are to be avoided.

2.21 When considering a component for manufacture by some form of casting, what are the principal factors that you would take into account in deciding whether casting was the most suitable method of production?

Chapter Three
Plastic deformation of metals

After reading this chapter you should understand:
   (a) what is meant by hot, cold and warm working of metals
   (b) how industrial steel sections are made by rolling
   (c) the principal methods and benefits of forging
   (d) the importance of the metallurgical effects caused by rolling and forging processes
   (e) the basic differences between rolling and extruding metal
   (f) what is meant by cold drawing
   (g) the way in which bar, wire and tubing are made, and how the mechanical handling problems involved are overcome
   (h) why components are made from sheet metal
   (i) the two fundamental processes involved in the production of most sheet metal parts
   (j) the benefits of fine blanking
   (k) the effect of applying shear to a punch or die
   (l) what press tools are and how they differ from transfer tooling
   (m) the basic difference between sheet metal bending and deep drawing
   (n) how spinning, shear and flow forming differ from other sheet metal processes.

3.1 The working of metal

Metals are said to be wrought when they have been subject to working, wrought being the past participle of the verb to work. The most common forms of metal working are rolling, forging, extrusion, drawing, spinning and shear forming, and in this chapter the principles of each of these processes are briefly described.

Working can be carried out either at ambient temperature (cold working) or at some higher temperature. How near a metal's melting point is to its working temperature determines whether working above workshop (ambient) temperature is classed as hot or warm working (Fig. 3.1).

3.1.1 Hot working

Metals worked above their recrystallization temperature (Van Vlack 1989) are said to be hot worked. A metal's recrystallization temperature is not quite as easy to determine as its melting

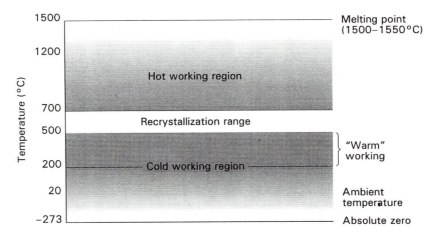

Figure 3.1 Temperature relationship between hot, cold and warm working of steels.

point, because it is influenced by such factors as the level of internal stress that exists in the metal before heating, the alloys contained within the metal and its melting point. For many metals the recrystallization temperature is between 0.4 and 0.5 of the material's absolute melting point (K). Depending upon carbon content, for most steels it is generally in the region of 500–700°C. As can be seen from Table 3.1, tin, zinc and lead have recrystallization temperatures not far removed from ambient, with aluminium being at just 150°C. Thus, hot working does not necessarily imply that the temperature of the metal being worked is above ambient, although in most cases it is.

An important characteristic of hot working processes is that the metal behaves in a perfectly plastic manner. Therefore the metal becomes neither internally stressed nor work hardened, so an unlimited amount of hot working can be performed on it without fear of component fracture.

Although the thermal energy costs involved in heating metals above their recrystallization temperature can be high, often this is more than outweighed by the large amount of reshaping that can be achieved in one operation compared with what is possible when working at a lower temperature.

Hot working causes slag (§2.4.1.4), porosity and other inclusions inherent to some degree in most cast metal to be deformed and broken up into insignificantly small pieces. Also, the coarse granular structure typical of cast metals is realigned and a large number of smaller

Table 3.1 Approximate melting and recrystallization temperatures for common engineering metals.

Metal	Melting point (°C)	Recrystallization temperature (°C)
Tin	230	−50
Lead	325	−50
Zinc	420	75
Aluminium	660	150
Copper (oxygen free)	1080	250
Cast irons	1150–1200	530–550
Steels	1500–1550	500–700
Titanium	1700	600
Tantalum	3000	1000

grains are formed. The effect of this is to impart a degree of toughness to the worked material, although hardness and density are unchanged by hot working.

Unfortunately, most hot working processes result in poor surface finish and produce items that are dimensionally imprecise owing to surface scale and the effects of uneven cooling.

3.1.2 Cold working

Cold working does not necessarily mean literally what it says! It is true that many metal working processes are carried out at ambient temperature, and these can certainly be termed cold working. However, metal working that is carried out at a temperature much higher than that prevailing on the shop floor can also be described as cold working (Fig. 3.1).

The upper temperature limit of cold working varies with the metal and its metallurgical condition before working, but for steels cold working is generally considered to be between ambient and about 300°C. Significant work hardening takes place, and this severely limits the amount of plastic deformation that can occur before the material ruptures. To overcome this, if all the desired geometric changes cannot be achieved in one operation, the material must be partly changed and then annealed to eliminate the stresses introduced by the work hardening. After annealing the workpiece can then be safely subjected to a second cold working operation that will, hopefully, permit reduction to the final desired geometry.

Cold working processes generally produce repeatable accurate components with a good surface finish.

Unfortunately, cold working forces are very much higher than those required for hot working, and unwanted residual stresses are usually present after the final stages of working. If they are significant they tend to be released during subsequent machining of the part, resulting in unpredictable distortion of the finished machined item. To avoid this, the part needs to be annealed before machining, to remove these locked-in stresses. Any distortion that occurs during annealing is of no consequence as its geometrical effects will, of course, be removed during machining of the part.

Note Recrystallization annealing is a softening process that is used to remove unwanted internal stresses existing in the microstructure of a cold-worked component. It involves controlled heating of the part in an oven to a preset temperature, holding that temperature, often for some hours, and then letting it cool at a carefully controlled rate. The typical stress-relieving temperature range for steels is 500–700°C, depending upon the steel's carbon content.

3.1.3 Warm working

Many texts take the simplistic approach that a metal worked below its recrystallization temperature is deemed to have been cold worked, even though this temperature may be much higher than ambient temperature. Although it is true that a metal worked below its recrystallization temperature no longer acts as a truly plastic material, there is an intermediate option, and this is called warm working.

Warm working, although not a totally stress-free process, is useful for working high yield strength, high-carbon steels. By heating the metal to a temperature between 0.3 and 0.5 of its absolute melting point, the deformation forces required are much reduced compared with cold working and consequently tool life is greatly improved. Some work hardening does occur, but accurate parts are produced with a surface finish approaching that obtainable by cold working.

3.2 Rolling

Almost all metals and their alloys begin life by being cast (Fig. 3.2).

Steel is the most common engineering metal and, with the exception of expensive special alloy steels, most steel made in Europe today is produced by continuous casting (§2.5.5). The rest is made from cast ingots (Fig. 3.2).

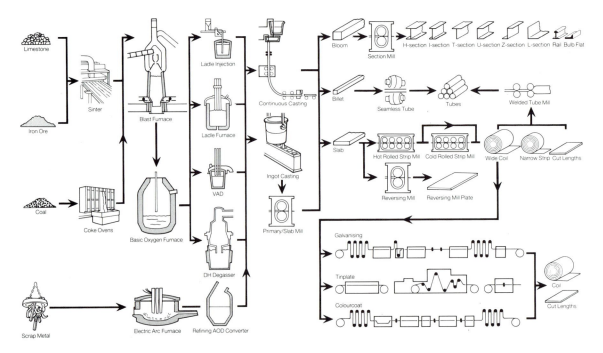

Figure 3.2   How steel is made from iron ore to semifinished product. (Courtesy of British Steel plc.)

Aluminium is the other commonly rolled metal, although many sections made from this metal and its alloys are formed by the extrusion process (§3.5).

Ideally the shape and size of the initial castings produced should be as near as possible to the form required by the end user, as this minimizes material and energy wastage; however, for practical reasons this is hardly ever possible. Therefore the as-cast metal usually requires reshaping, the first reshaping operation normally being rolling.

The most usable shapes produced by the primary rolling process are slabs (thick, flat plate), blooms (large rectangular bar), billets (large square bar), structural sections, tube and rail section. Secondary rolling is used to produce sheet metal, and small-section round, square and hexagonal bar.

Exercise   You are asked to make some steel bolts 15 mm diameter × 75 mm long. Would you prefer to have the material to make these delivered to your lathe in the form of hexagonal or round bar? Why?

The metallurgical and physical properties required of the finished material determine whether hot and/or cold rolling is used, although the basic principle involved is the same.

3.2.1 Hot rolling

Irrespective of whether the as-cast metal is in the form of slabs produced by continuous casting or ingots weighing up to 6 tonnes, made by traditional mould casting, it will require size reduction and reshaping to be of any practical use. This is achieved by passing it through a pair of rotating cylindrical steel rollers, which impose a squeezing action on the metal and reduce its thickness appreciably.

(a)

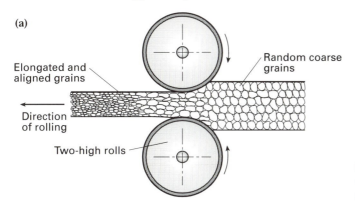

Elongated and aligned grains

Random coarse grains

Direction of rolling

Two-high rolls

Figure 3.3 (a) Grain flow in direction of rolling on a two-high cogging mill. (b) Four-high mill stand. (Courtesy of British Steel plc.)

(b)

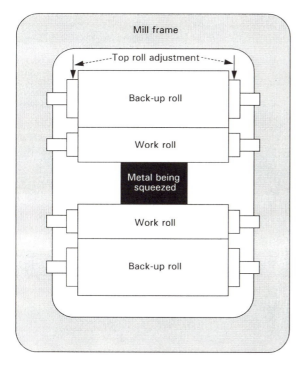

Mill frame

Top roll adjustment

Back-up roll

Work roll

Metal being squeezed

Work roll

Back-up roll

The overall reduction in thickness required can be very large, so to avoid the need for an excessive number of squeezing steps, or *passes*, primary rolling is carried out hot (§3.1.1). Even so, very large reductions, such as reducing a square 650 mm ingot down to a square billet below 200 mm, can still require a large number of passes. This is because of the practical limitations on the force that can be applied by the rolls – up to 1000 tonnes in some cases. This, in turn, limits the reduction per pass to about 50 mm when rolling steel.

Similar limitations apply to the thick rectangular slabs produced by continuous casting; these are usually subject to secondary rolling to produce rectangular plate and thin strip.

The primary rolls are termed a cogging mill and, in addition to drastically reducing the metal's thickness, also change its coarse, as-cast, granular structure to a more refined and elongated form that is aligned in the direction of rolling (Fig. 3.3a).

The terms "two high" and "four high" are used to describe the number of rollers in the mill frame, depending on whether or not the two rollers in contact with the work material (the work rolls) are supported by a pair of large reinforcing back-up rollers to minimize their deflection during passage of the metal through the mill frame (Fig. 3.3b).

Since the volume of an ingot or slab remains constant during rolling, length increases as thickness reduces during successive passes, resulting in exit speeds of blooms approaching 40 mph. This, coupled with the high temperature of the metal, approximately 1200°C, makes steel rolling an extremely dangerous process, requiring highly skilled teams to operate such plant safely.

Additional hot rolling operations are used to reduce stock thicknesses still further and/or to produce specific profiles such as rail section, I-beam, channel, angle, bar and strip. To generate these profiles the work rolls have the required shape cut into them in a series of progressive steps. The feedstock is passed sequentially through these specially shaped rolls, emerging from the final pass with the required form (Fig. 3.4).

The four-high rolls used for this type of work are often placed in series with the primary rolls, which can result in a rolling mill plant up to 1 km long. Alternatively, a separate rolling line may be used, but this usually means that the blooms or slabs need to be reheated to rolling temperature. Since over 80 per cent of the total energy used by a rolling mill is consumed in reheating (British Steel 1992), there are strong financial incentives to go from ingot or slab to finished section in one continuous operation.

Large quantities of water are used in hot rolling, principally to keep the mill rolls cool and to flush away the surface scale formed as the hot metal cools during rolling. This prevents the scale from being rolled back into the surface of the material.

Exercise   With such large quantities of water being sprayed on the mill rolls, why does this not chill the metal passing through them?
Hint Compare the densities and specific heat at constant pressure values for water and steel.

When producing strip (continuous thin sheet), the rolled plate is normally passed through finishing rolls, which outputs a thickness of 2–3 mm, metal temperature having by then fallen to almost half of its value at the beginning of rolling. Finishing rolls in excess of 3.5 m wide are now available.

3.2.2 Cold rolling

Cold rolling is a post-hot rolling operation and is used only when the metallurgical and dimensional properties (such as straightness) of the final product require it. The process is the same as that used in hot rolling, except that, despite the limited size reduction per pass possible when working below the recrystallization temperature of the metal, only four-high mills (Fig. 3.3b) are used because the forces imposed on the working rolls are extremely high.

Pass:-   1   2   3   4   5   6   7   8   8(again)

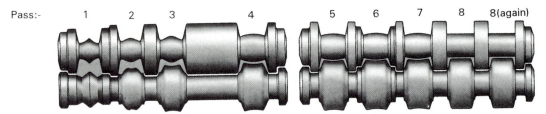

Figure 3.4   Progressive profile rolling of a channel section. (Courtesy of British Steel plc.)

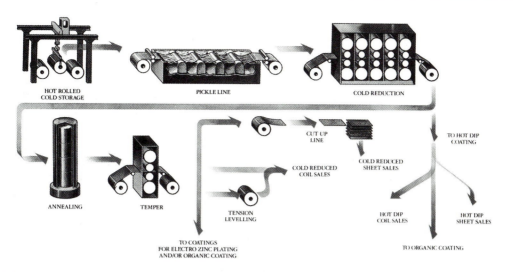

Figure 3.5   Cold rolling of strip steel. (Courtesy of British Steel plc.)

Before cold rolling can be carried out it is necessary to remove any surface scale left after hot rolling. This is achieved by immersing the metal in an acid pickling bath – for steel hydrochloric acid is used. All traces of acid are then washed off with water, the steel is dried with hot air and finally coated with a thin film of oil to prevent surface corrosion. It is then ready for cold rolling.

Cold rolling increases metal toughness and provides it with a degree of surface hardness because of the work hardening that results from cold working. While these properties can be advantageous, work hardening severely limits the amount of reduction that is possible in each rolling pass if metal cracking is to be avoided. Annealing (§3.1.2) is therefore normally performed either between rolling passes, depending upon the number of passes required, or after final rolling. For steel, annealing is usually carried out in an inert atmosphere to avoid surface oxidation.

The principal steel products that are cold rolled are round, hexagonal, square and rectangular bar (known as bright bar) and sheet/strip.

When cold rolling sheet metal, where thickness accuracy and surface finish are critical, a final post-annealing operation is usually carried out. This is yet another rolling pass but involves a thickness reduction of only about 1 per cent, and provides a controlled degree of surface hardness. This is called surface tempering and produces strip material of exceptional dimensional accuracy and surface finish. It also ensures the establishment of the metallurgical properties required for subsequent processing by the end user.

In modern rolling plants, if the strip material requires surface coatings (Ch. 6), the equipment needed to apply them is frequently incorporated within the flow path through the mill (Fig. 3.5).

3.3 Forging

In contrast to rolling, the end product of which is long lengths of metal deformed into the desired shape by the continuous application of compressive force, forging produces individual components as near to the final shape as possible.

49

Forging is similar to rolling in that the transformation of a block of raw material (a billet) into the required component shape is achieved by the application of an external force. However, with forging this is normally in the form of an intermittent compressive or impact force, although closed-die pressure forgings (§3.3.3) are made by applying a continuous compressive force with a hydraulic press.

Although forging is generally a hot working process, in Japan and Germany, cold forging is becoming increasingly important, particularly in the automotive industry. An example of this is cold upset forging (§3.3.4).

A most important feature of forging is that the components produced have greatly enhanced impact strength and toughness compared with the same components produced by alternative methods of manufacture such as casting or machining from the solid. As a wide range of metals can be forged, this explains the popularity of the process, particularly for the production of highly stressed components such as crankshafts for racing car engines.

3.3.1 Hammer forging

The simplest form of hammer forging is hand forging, which is the process used by the village blacksmith.

The process is simple. First the workpiece material is heated in an open fire until it is well above its recrystallization temperature (for steels to approximately 1250°C). It is then placed on a specially shaped steel work surface called an anvil and beaten into the required shape with a hammer. Considerable experience is needed to assess when the metal is at a sufficiently high temperature, as well as to know where and how to hit the metal to produce the required shape. Nevertheless it is a very cost-effective way of making small quantities of moderately sized components.

One disadvantage of blacksmiths is their physical strength and stamina limitations compared with mechanical hammering devices. This is why most commercial hammer forging is carried out using large pneumatically or mechanically driven hammers, although gravitational drop hammers are also still used in many forging shops.

Even with the much greater force available from a power hammer, the forgemaster still needs a great deal of skill to produce the required shape. However, if large numbers of identical components are required, much of this artisan skill can be eliminated by the use of closed-die drop forging.

Exercise   The force applied to a workpiece in impact forging is derived from the kinetic energy imparted by the forging hammer, and is simply calculated from the equation $\text{KE} = \frac{1}{2}Mv^2$, where M is the mass of the hammer and v is the velocity of the hammer at the instant of impact with the workpiece. To calculate v for a hammer in gravitational free fall, you need to recall the basic equations of motion. Do this, and then attempt question 3.10 at the end of this chapter.

3.3.2 Closed-die drop forging

Drop forging, or hot stamping as it is sometimes called, is identical in principle to hammer forging except that it places less reliance on the skill of the forgemaster to produce the precise shape required. Instead, a pair of mating steel dies (a die-set) are used; when brought together, these form an enclosed cavity that corresponds to the shape of the required forged component (Fig. 3.6). One half of the die-set is attached to the drop hammer and the other half-die is secured to the anvil.

A billet is heated above its recrystallization temperature and then placed on the surface of the lower half of the die-set attached to the anvil. The upper die is then either allowed to free fall or is forced down by pneumatic or mechanical means onto the bottom die, trapping the hot metal within the closed die-set cavity. This forces the hot metal to deform and fill the cavity, so producing a repeatable and accurate forging. In practice, one blow is rarely sufficient to fill the die cavity completely, so a number of blows or "drops" are normally required.

Drop forging is a flexible process: to produce a different component only the die-set has to be changed. Figure 3.7 illustrates, from left to right, how a pair of engine connecting rods are forged from a length of round bar in six steps. Finally a trimming operation (§.3.7.1.1) is required to remove excess metal (the flashing) from around the periphery of the finished part. (The scrap flashing is shown in the last photograph of Figure 3.7.)

When metal is removed from a furnace it starts to cool immediately, and if forging cannot be completed before the temperature falls below the metal's recrystallization temperature then the unfinished forging must be reheated to prevent cracking problems. As cooling is time dependent, there is a powerful commercial incentive to complete forging quickly and avoid the need to reheat the unfinished part to complete the job. With closed-die forging reheating is rarely necessary, except for the largest forgings, although it is common practice with open-die (hammer)

Figure 3.6 Typical forging die-set for making engine connecting rods. (Courtesy of Eumuco aG, Leverkusen, Germany.)

Figure 3.7 Stages in drop forging connecting rods using the die-set shown in Figure 3.6. (Courtesy of Eumuco aG, Leverkusen, Germany.)

forging. Reheating is not required in the closed-die pressure forging process (§3.3.3) because it produces parts with just one stroke of the press.

3.3.3 Closed-die pressure forging

Unlike hammer and drop forging, closed-die pressure forging does not rely upon kinetic energy from repeated hammer blows to achieve the required metal deformation. In this process a steady and controlled pressure, typically 1000–3000 tonnes, is applied to the hot billet, squeezing the metal into the shape of the cavity within the die-set. This is done in one operation and is therefore quicker and cheaper than drop forging. Although a press forge is much more expensive than a drop forge, its greater output, improved product accuracy and suitability for process automation more than compensate for the additional capital cost.

Press forging also produces a metallurgically superior product to drop forging in that the billet is subject to hot working throughout its whole thickness, whereas with impact forging the microstructure benefits of hot working are confined to the component's surface region.

Closed-die press forging is also used to forge components from powders rather than from metal billets. This is dealt with in Chapter 7 (§7.2.5.1).

3.3.4 Upset forging

In upset forging the cross-section of a billet, rod or bar is increased, and its length correspondingly reduced, i.e. the volume of the stock remains the same (Fig. 3.8). The process involves applying lengthwise impact pressure to one end of the blank, and forging it into the required shape.

This process is particularly useful for large-quantity production of bolts: upset forging of the hexagonal head avoids the wastage of raw material and time that occurs when they are made from hexagonal bar. Engine inlet and exhaust valves and rivets are other good examples of components made by upset forging. However, there are certain practical limitations; for example, to avoid the risk of the workpiece material buckling under the axial forging force, the

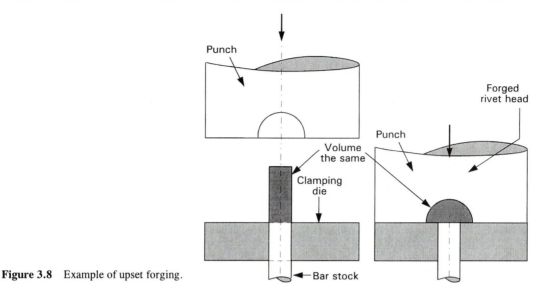

Figure 3.8   Example of upset forging.

maximum length of unsupported material that can be upset in one blow must not exceed about 2.5 times its diameter.

Exercise   Think of other suitably shaped items that are made in large quantities and which lend themselves to upset forging. The greater the material saving, the more attractive commercially this process becomes.

Hint   Do not limit your thinking to parts that have an enlarged end.

Although upsetting is generally a hot working process, it can also be carried out cold providing the degree of deformation required is not too great. Also, surface finish and component accuracy are usually superior to that achieved by the hot forging process.

A good example of the application of this type of cold forging is the production of carpentry nails whose heads are upset forged. This is also known as cold heading. Yet another application is the first forming stage in the production of steel bodies of sparking plugs from circular, low-carbon steel bar stock (Fig. 3.9).

With items produced in such large quantities the whole process cycle from bar stock to end product is generally fully automated, giving extremely short cycle times.

3.3.5 Roll forging

When short, simple shapes are required, it is possible to combine the principle of rolling with that of press forging. This process is known as roll forging and involves the use of a pair of cylindrical rollers that have the shape of the required component cut into their surface – similar to progressive rolls used in continuous rolling and shown in Figure 3.4. However, unlike progressive rolls, the pair of rolls used in roll forming are shaped to the complete three-dimensional profile of the forged part over just half of its circumferential distance (πD). Because the component's shape is wrapped around the periphery of the pair of rolls this process

Figure 3.9 Stages in forging and extruding the body of a sparking plug. (Courtesy of Champion Spark Plug Co., Wirral, Merseyside.)

is only suitable for the production of short components if the diameter (*D*) of the rolls is to be kept within practical limits.

The forging operation involves a heated bar of feedstock being fed into the rolls, which are then rotated for one-half of a revolution, at the end of which the forged part is ejected (Fig. 3.10).

Depending upon component complexity, more than one pass may be required, the profile for each pass being cut sequentially over the width of the rolls, as shown in Figure 3.11.

With the exception of items having relatively simple shapes, such as garden spades, most roll forging is carried out as a preforming operation before drop forging. Figure 3.12 illustrates this principle as applied to the first stages of production of the engine connecting rods shown in Figure 3.7. Because two rods are produced per operation, the production rate in this example is 1400 rods per hour.

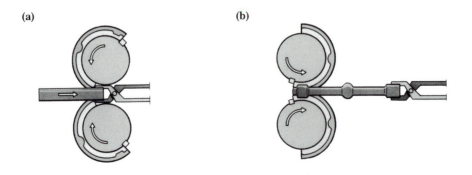

Figure 3.10 Principle of roll forging. (Courtesy of Eumuco aG, Leverkusen, Germany.)

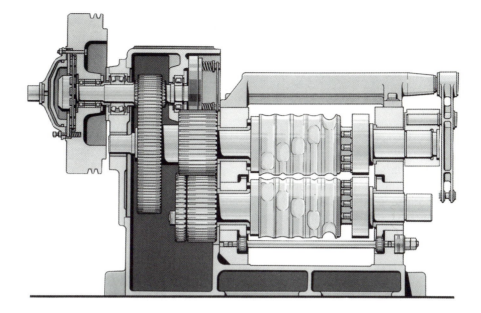

Figure 3.11   Progressive forging rolls. (Courtesy of Eumuco aG, Leverkusen, Germany.)

Figure 3.12   Forging rolls for pre-drop forging operation. (Courtesy of Eumuco aG, Leverkusen, Germany.)

3.4 Principal metallurgical effects of rolling and forging

Both rolling and forging produce a "grain" flow pattern in the material similar to the grain in wood. This results in mechanical properties varying with grain direction (called *anisotropy*). Figure 3.13 illustrates how the grain flow pattern of part of an engine crankshaft varies when the method of production is by forging, machined from solid rolled bar, and cast.

Forging produces a particularly useful grain pattern in that it tends to follow the physical shape of the component (Fig. 3.14). This imparts increased strength in the same areas that the highest stresses are likely to occur in use, e.g. at corners and section changes. This is why high-performance engine crankshafts are forged rather than cast. When machined from the solid the grain flow follows the direction of rolling (Fig. 3.3a); in contrast, the granular structure of cast parts is largely random. Thus, when designing a critically stressed component, it may be possible to minimize the risk of in-service failure either by specifying a forging rather than a casting or, if produced from rolled bar or sheet, by ensuring that the component's direction of grain flow is such that maximum strength coincides with the direction of maximum loading.

Other advantages of the rolling and forging processes are:
- Steel forgings are particularly weldable.
- There is little waste material in forging, and minimal scrap with closed-die forging.
- Any slag inclusions in the material before rolling and forging tend to be broken up into insignificantly small pieces and dispersed within the metal during these working processes.
- Work hardening, particularly of the surface, occurs during cold rolling and forging. The resulting hard skin makes subsequent machining more difficult, although it can be advantageous as the resulting enhanced material toughness and surface hardness may enable the designer to specify a lower (cheaper) grade of metal than would otherwise be possible. A degree of work hardening also improves a metal's impact resistance.

Forging   Machined from Bar stock   Casting

Figure 3.13   Grain flow patterns in an engine crankshaft. (Courtesy of British Forging Industry Association, Birmingham.)

3.5 Extrusion

Extrusion is the process of exerting sufficient compressive pressure on a ductile slug/billet of metal to force it to flow out through a hole in a steel die of the required shape, like squeezing a household adhesive from its plastic container. It may be carried out either hot or cold, i.e. above or below the recrystallization temperature. Although steels can be extruded, the ram force required is high, even when hot extruding. Therefore extrusion tends to be limited to more ductile non-ferrous metals such as aluminium, magnesium, zinc and copper alloys.

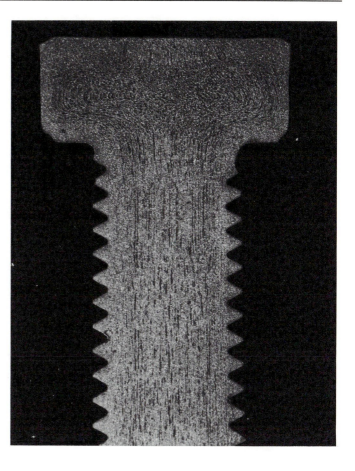

Figure 3.14 Grain flow macrograph of a forged high strength bolt. (Courtesy of ERA Technology Ltd, Leatherhead.)

Long lengths of the most complex shape are easily extruded (Fig. 3.15), and the shape produced can be quickly altered simply by changing the extruding orifice die. This makes even small batches economical.

Exercise Look for everyday examples of complex two-dimensional aluminium extrusions. Examine them closely and note the quality of finish achieved. You should also be able to see fine longitudinal marks that were formed as the metal passed through the die.
Hint Do not confine your search to industrial applications.

There are three main forms of extrusion: direct, indirect and impact. The first two are usually carried out hot, with impact extrusion being generally a cold working process.

3.5.1 Direct extrusion

A billet of metal is placed in a cylindrical chamber fitted at one end with a die whose orifice profile corresponds to the cross-section of the required extrusion. A ram then forces the hot

Figure 3.15   Typical complex extruded aluminium profiles. (Courtesy of Hydro Aluminium Ltd, Warwick.)

plasticized metal out through the die to produce a length of the required shape (Fig. 3.16). Obviously, the larger the billet of material used the longer will be the extrusion.

In addition to producing lengths of complex solid two-dimensional shapes, extrusion can also be used to make hollow sections by inserting a mandrel through the centre of the ram.

3.5.2 Indirect extrusion

Indirect extrusion is similar to the direct process except that the extruded section is forced back through the centre of the ram (Fig. 3.17). While this may appear to complicate the extrusion process unnecessarily, less ram force is required as there is no frictional resistance between billet and chamber wall because there is no relative motion between them.

Nevertheless, direct extrusion is more popular because it is easier to support the extruded section adequately as it emerges from the die, and makes longer extrusion lengths more practicable. Also, a solid ram is easier to make, and is more rigid and stronger than the hollow alternative.

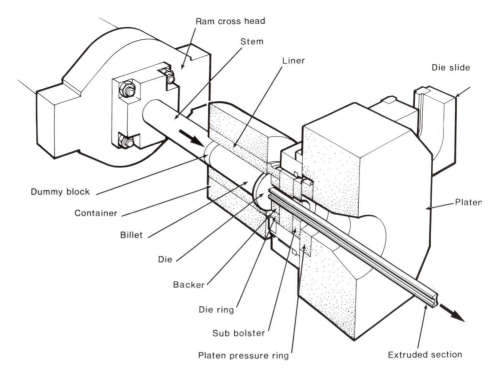

Figure 3.16 Circular billet extruded into an I-section. (Courtesy of UK Aluminium Extruders Association, Birmingham.)

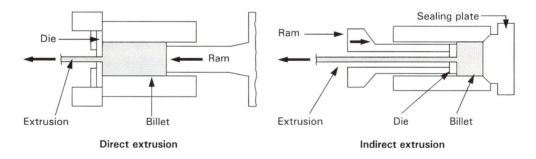

Figure 3.17 Difference between direct and indirect extrusion. (Courtesy of UK Aluminium Extruders Association, Birmingham.)

Exercise Do not overlook the fact that metals such as lead, tin, and aluminium all have recrystallization temperatures around ambient, so hot extrusion could be carried out at or near shop floor temperature. Why to you think billets are nevertheless normally preheated, for example aluminium is usually heated to 450–500°C before extrusion (UK Aluminium Extruders Association 1991)?

3.5.3 Impact extrusion

This extrusion variant involves striking a slug of metal with a punch using sufficient force that the metal in the die fills the annular gap between punch and die. In this respect the process is similar to closed-die drop forging. It is most frequently a cold working process, although the impact speed is such that a significant temperature rise occurs within the billet at the time of extrusion. It is suggested that this rapid temperature rise is the reason why it is possible to achieve a much greater degree of deformation with this process than would normally be expected when cold working.

Impact extrusion clearly lends itself particularly well to automated high-speed production of thin tubular components in non-ferrous materials such as aluminium and zinc. Cycle times of less than 1 second are common when extruding such metals. Impact extrusion is also used to mass produce certain components made from ductile low-carbon steel. A good example is the manufacture of sparking plug bodies (Fig. 3.9) where all forming stages, other than the initial upset forged one (§3.3.4), are carried out by impact extrusion.

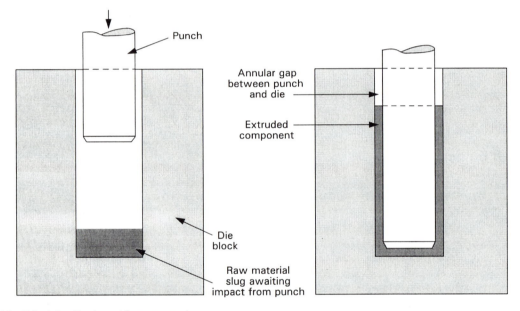

Figure 3.18   Principle of backward impact extrusion.

3.5.4 Some advantages and limitations of the extrusion process

Steel is difficult to extrude (particularly impact), owing to its high strength even when hot worked. Hot working at temperatures of 1000°C or more presents numerous practical problems, such as a tendency for the steel billet to semiweld itself to the steel die and extrusion chamber, although this can be overcome by using graphite or glass as a lubricant. The melting point of glass is just below that of steel, but frictional heating between the glass and the hot steel billet is sufficient to provide a steady film of molten glass lubrication.

The ratio of the cross-sectional area of the billet and the extruded shape is known as the extrusion ratio (ER). For aluminium the optimum range of ER when hot extruding is between 30 and 50 (UK Aluminium Extruders Association 1991), but for steels this figure is much lower owing to the greater forces imposed upon the extrusion press and die – up to 20 times greater in some cases.

In general, cold extrusion is limited to the softer metals to avoid the need for large expensive presses. Also, length/diameter ratio is normally a major limiting factor in cold extrusion, maximum values being about 10:1 for aluminium but only 3:1 for even the most ductile steels.

In certain instances cold extrusion can be competitive with deep drawing (§3.7.2), as less expensive tooling is required; the production of dry cell battery cases is a good example of such a situation.

Extruded parts are often much lighter and stronger than their cast or forged equivalents, tolerances are closer and less subsequent machining (if any) is necessary.

3.6 Cold drawing

Cold drawing is the term applied to a range of cold working finishing processes involving either thin rod, wire, extruded tube or sheet metal. Cold drawing of sheet metal is usually referred to as deep drawing, and is discussed in section 3.7.2.

Cold drawing of rod, wire and tube may be considered to be the opposite of direct extrusion in that material stock is pulled through a die rather than pushed through it. As with all cold working processes, work hardening occurs and the grain flow in the finished product runs in the direction of working.

3.6.1 Wire drawing

Industry uses large quantities of thin rod and drawn wire because they are clean, dimensionally accurate and have a good surface finish. Coiled wire is also a convenient form of raw feedstock.

While bar/rod is produced by cold rolling, circular wire is produced by cold drawing and involves pulling material through a tapered die to reduce its diameter from D_1 to D_2 (Fig. 3.19).

As this is a cold working process there is a limit to the amount of reduction possible before the material work hardens to such an extent that it fails in tension. Nevertheless, up to a 90 per cent overall reduction is possible with a multipass system without annealing providing the low-

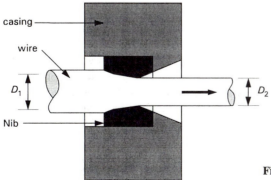

casing

wire

D_1

D_2

Nib

Figure 3.19 Wire drawing die unit.

61

carbon steel used is fully annealed before drawing commences. If very fine wire diameters are required interstage annealing is often necessary. The most common wire diameters range from 2 to 4 mm, but wires can be drawn down to as little as 0.25 mm diameter.

Because several serial drawing stages are usually necessary, and because a very long length of wire is produced from only a moderate length of stock, mechanical handling becomes increasingly difficult as each stage progresses. As the stock diameter reduces so the drawn length increases proportionally with the square of diameter reduction. To overcome this difficulty the wire is wound onto an accumulator block (storage drum) after each pass (Fig. 3.20). These accumulator blocks enable very high drawing speeds to be used – up to 1500 m/min (almost 60 mph).

In practice, most wire drawing machines consist of a series of progressively reducing dies, with an average diameter reduction per die of 20–25 per cent (Fig. 3.21).

Figure 3.20   Die and accumulator block for cold drawing of wire. (Courtesy of Tinsley Wire Co., Sheffield.)

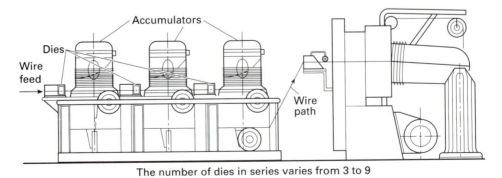

The number of dies in series varies from 3 to 9

Figure 3.21   Typical multistage wire drawing system. (Courtesy of Tinsley Wire Co., Sheffield.).

Exercise   To get a feel of the magnitude of the mechanical handling problem when wire drawing, calculate the length of 5 mm diameter wire produced from feedstock 22 mm diameter and 10 m long. (Remember that the wire is probably being produced at a speed of between 40 and 60 mph!)

3.6.2 Tube drawing

Tube drawing is also similar to thin rod and wire drawing, with the exception that a fixed or fully floating central plug is used to produce the required central hole (Fig. 3.22).

Tube is generally produced from a hot-extruded annular-shaped billet. This is then drawn down to the required size and shape (not all tubes are circular) in a number of stages, with interstage annealing when required. Much of the tubing produced is made from copper and its alloys, with external diameters from 400 mm down to 0.5 mm being common.

Small-diameter tubing, like wire, also presents a mechanical handling problem. Although this difficulty is again solved by the use of drum storage, extra care must be taken to ensure that the finished product is not too tightly coiled to avoid the central hole being deformed.

Larger diameters are produced in relatively long lengths (Fig. 3.23) and then cut into more manageable lengths of approximately 10 m.

The cold drawing process is particularly attractive to tubing producers as it offers exceptional surface finish as well as consistent accuracy of wall thickness, bore and outside diameter. It is also possible to achieve the required degree of surface hardness and material toughness by suitably adjusting the number of draws used, the degree of reduction made on the final draw and the final heat treatment.

Exercise   Over the next week list as many uses of metal tubing as you can think of, making a note of the material used in each case. Why do you think the material selected was chosen? Also, closely examine the surface finish, particularly of copper tubing, and see if you can find evidence of cold drawing.

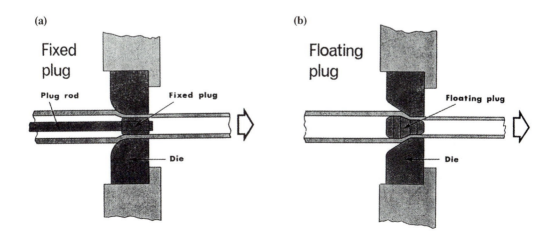

Figure 3.22   Fixed and floating plug tube drawing. (Courtesy of IMI Yorkshire Alloys Ltd, Leeds.)

63

Figure 3.23 Drawing of large-bore copper tubing. (Courtesy of IMI Yorkshire Alloys Ltd, Leeds.)

3.7 Sheet metal component manufacture

The production of items from sheet metal (usually in annealed low-carbon steel) constitutes a significant sector of manufacturing industry.

This process is used when the alternatives, casting, machining from the solid, forging, etc., are all too expensive in terms of time and materials or are impracticable. Imagine making car body panels or washing machine and refrigerator outer casings in any other way than from sheet metal!

The process involved in making parts from sheet metal readily lends itself to mass production, which is fortunate, as the tooling used in cutting and bending is usually expensive. Nevertheless, the parts produced normally have a good finish, little material is wasted and, if correctly designed, the fabricated products, as they are called, can offer a high level of strength and rigidity.

3.7.1 Sheet metal cutting and bending

To produce parts from sheet metal usually involves two basic processes:
1. shearing (cutting) the sheet material to the correct size and shape
2. bending it to the required shape, frequently in a number of stages if the shape is complicated.

Each of these two activities is now briefly examined.

3.7.1.1 Shearing

Most sheet metal parts are subject to shearing (note that blanking, piercing and trimming are all forms of shearing) at some stage in their manufacture. This can range from cutting out an initial blank to trimming the final product to size and may include the piercing of holes at various points. However, the fundamental process of shearing is always the same and is similar to the cutting action of a pair of domestic scissors if one blade was held fixed and the other one moved relative to it.

Shearing is really a controlled fracturing process along a precisely defined line, and is achieved in the following way. A force is applied along the required cutting line by the cutting

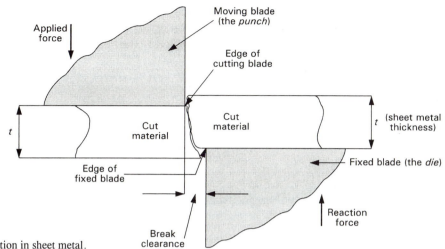

Figure 3.24   Shearing action in sheet metal.

blade (referred to as a punch), and this produces shear stress in the material. As the shearing force is increased so shear stress increases along the cutting line, causing plastic deformation to occur to such an extent that the metal's ultimate shear stress (σ_s) is exceeded and the material finally fractures or breaks along the cutting line (Fig. 3.24).

Obviously, the shape cut will not always be a straight line, one of the most commonly produced shapes being a hole. Holes can have any contour, although by far the most common shape is a circle, i.e. a round hole. The principle of shearing a hole is identical to cutting in a straight line, and can be considered to be the process of cutting along a line that curves through 360°.

A small clearance, called break clearance (BC), must be provided between punch and die. Its size is largely a function of the ductility of the metal being cut, and its thickness, and varies from about 2 per cent of material thickness for a ductile material such as aluminium to 10 per cent of material thickness for a medium-carbon steel. If insufficient break clearance is provided unnecessary cutting work is required, *although it has no effect on the peak shearing force (F_{max}) required*, as this is purely a function of the material's ultimate shear stress (Fig. 3.25). If too much BC is provided the metal bends along the shearing line before fracture occurs, and this causes an unacceptable curved edge on the part.

If, during cutting, the material that drops through the die (the slug) is unwanted scrap, then the process is termed a punching operation. If the metal that goes through the die is the required part, then this is called blanking. In either case the cutting mechanism involved is the same.

A simple example of the use of these two operations is the manufacture of plain washers – the central hole is punched out and the outside diameter of the washer is blanked out from the appropriate thickness of strip material.

When punching holes, BC is required all

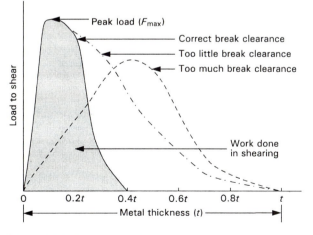

Figure 3.25   Effect of break clearance on shearing work required.

around the punch, thus it is necessary to allow for two BCs when calculating the diameter of either punch or die. Whether the BC should be applied to the punch or die is simple to deduce logically – always a better learning approach than trying to rely on memory. When punching a hole its size is determined by the size of the punch, i.e. in the case of a round hole the punch diameter must be the same size as the required hole. Therefore two BCs must be added to the diameter of the hole in the supporting die. In blanking operations the hole through which the component is forced determines the size of the part produced. Therefore, when blanking circular items two BCs must be deducted from the diameter of the punch, since its size is not critical as it only serves to push the metal part through the hole in the die.

Exercise "Trimming" is the shearing process used to remove the unwanted material around a finished component. Flashing around forgings (Fig. 3.7) is a good example of this, but would you say that trimming in this context is really a blanking or punching type of operation?

One problem with cutting sheet metal as shown in Figure 3.24 is that, even with the correct BC, the outline produced needs to be deburred, i.e. the sharp edges must be removed. To avoid this the process of fine blanking was developed.

Fine blanking is more akin to extrusion than shearing. The material is both rigidly clamped by a gripper ring (which slightly bites into the metal to ensure that it does not move during cutting) and supported from underneath throughout the actual cutting operation by a cushioning ram (Fig. 3.26). Although this technique produces accurate, burr-free parts, the press required is complex, and its cost can only be justified if production requirements are large. In addition, fully annealed fine-grained strip must be used if the best results are to be achieved.

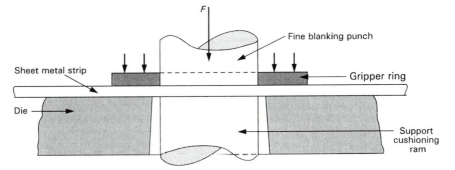

Figure 3.26 Principle of fine blanking.

Judicious planning to guarantee the efficient utilization of material is well worthwhile as most scrap metal is generally worth only 50 per cent or less of the cost of the original material. When a wide range of shapes are required it is possible to enlist the help of special computer programs to optimize the nesting of pieces so that optimum material utilization is achieved (Fig. 3.27). Utilization figures of 97–98 per cent are now achieved with computer assistance, particularly in the aerospace industry, where the metals used are often extremely expensive. However, to achieve such high utilization figures requires laser cutting (§8.9.4.3) rather than the shearing process described here.

To establish the required press size for any given job, the maximum shearing force (F_{max}) can be calculated from:

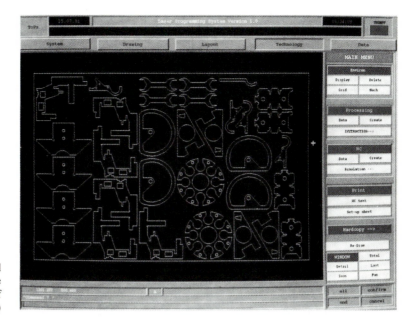

Figure 3.27 Computerized component nesting to maximize material utilization. (Courtesy of Trumpf Ltd, Luton.)

$$F_{max} = t \times \text{perimeter of profile to be cut} \times \sigma_s$$

where t is the material thickness and σ_s is the ultimate shear stress of the material.

In the case of annealed low-carbon steel, ultimate shear stress is approximately 75 per cent of ultimate tensile stress σ_{ult} (a more commonly quoted value). For this metal the above equation can therefore be approximated to:

$$F_{max} = t \times \text{perimeter of profile to be cut} \times 0.75\sigma_{ult}$$

When cutting holes of diameter D, the perimeter of the profile cut is equal to πD, and the above equation applicable to annealed low-carbon steel becomes:

$$F_{max} = t\pi D \times 0.75\sigma_{ult} = 2.4\, tD\sigma_{ult}$$

It is possible to reduce the above force (F_{max}), and hence the press size required, by chamfering either the punch or the die. This is called applying "shear" (Fig. 3.28), and has the effect of making the cutting process more progressive.

When blanking, the punch should be flat so that the part being pushed through the die remains flat; thus shear is applied to the die. When punching, the slug pushed through the die is scrap, so it does not matter whether it is bent or not; thus shear is normally applied to the punch in punching operations.

If too much shear is applied, the punch becomes fragile and prone to breakage, so a compromise has to be reached between minimum shear force and reasonable tool life.

With shear (S) applied, the reduced press force necessary (F) is given by:

$$S = \frac{\left[(F_{max} - F) \times t \times \text{per cent penetration into material at instant of shear}\right]}{F}$$

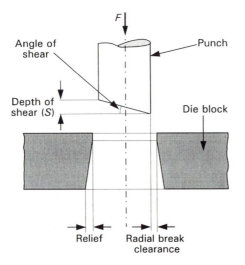

Figure 3.28 Single shear applied to the punch.

Low-carbon annealed steel shears when punch penetration reaches approximately 40 per cent of the material thickness, but if it becomes work hardened this figure is nearer 30 per cent Although this enhances the beneficial effect of applying shear it is more than offset by the increase in F_{max} that results from work hardening. Because bending causes work hardening, it therefore makes sense to carry out as many metal cutting operations as possible before bending, which fortunately is the most logical procedure anyway.

Exercise   Examine an ordinary office paper punch. You should find evidence of shear being applied to the punch pins to minimize the force needed to pierce a number of sheets at once, but why do you think some punch designers use "double shear" on each pin rather than a single shear angle as shown in Figure 3.28?

3.7.1.2 Bending

Because most sheet metal parts are subject to bending operation(s) while being turned into the required shape, it is essential to understand what occurs during the basic bending process. When sheet metal is bent, one side is subjected to tensile stresses and the other side to compressive stresses (Fig. 3.29), but as long as these stresses are well within the safe stress limits of the material there is no problem. In the context of bending, safe stress limits are clearly very different to those used under normal operational design conditions. i.e. well below the elastic limit of the material. Indeed, if the metal is not stressed beyond the elastic limit during bending, no permanent deformation will occur and so no new shape will be produced. However, it is obviously not possible to go too near to the ultimate tensile stress of the material if one wishes to avoid a high level of scrap.

Thus, it should be understood that sheet metal bending is a plastic deformation process and not one carried out within the elastic working region of the metal. This is why algebraic relationships applicable to elastic region working, such as Euler bending and simple beam theory, cannot be applied to sheet metal bending. Indeed, experimental tests have shown that calculations based upon elastic theory to determine press loads often underestimate the true forces

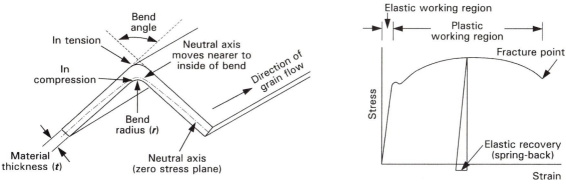

Figure 3.29 Simple bending of sheet metal.

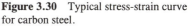

Figure 3.30 Typical stress-strain curve for carbon steel.

required by more than 100 per cent. As is so often the case in engineering, there is no substitute for experience backed up by practical tests.

An important factor in bending is the tightness of the bend produced, normally defined as the radius/material thickness (r/t) ratio. The more acute the bend, the greater are the stresses generated in the metal and the greater the risk of exceeding the material's ultimate stress limits in the bend area. Furthermore, with tight (low r/t) bends, thinning of the workpiece material occurs, aggravating still further the local high stressing problem. For example, typical percentage reductions in thickness of a fully annealed low-carbon steel sheet, as a function of bend radius, are 10 per cent for $r/t = 4$ but 25 per cent for $r/t = 1$.

The minimum bend radius for any given application is largely a function of the ductility and stress conditions within the material used and, surprisingly, the smoothness of the edges of the component blank.

When the load is removed from metals stretched beyond their elastic limit in bending, the metal releases the elastic strain energy needed at the start of the bending process to raise it to its elastic limit. This results in the component springing back slightly from the angle to which it has been bent. This phenomenon is called springback (Fig. 3.30), and while it is numerically small (typically 1–2° in low-carbon steel) it still has to be accounted for in the design of the tooling. This is usually achieved by overbending slightly beyond the required angle. The r/t ratio also has a major effect on the level of springback that occurs: the larger the bend radius the greater the amount of springback.

3.7.1.3 Press tools

A key to successful sheet metal cutting and bending is the design and manufacturing quality of the tooling used – referred to as press tooling. Press tools can be used for single operations, such as blanking, punching and bending, but if the production quantities merit it a combination press tool may be justified. This is simply a press tool that incorporates a number of single tools so that, at each stroke of the press, an operation is carried out simultaneously at each station (Fig. 3.31). This can result in a completed part at the end of each press stroke if it is not too complex. If the component is too elaborate to be finished in one combination press tool, it will need to be transferred to one or more other tool stations, and if transfer is performed automatically the whole installation is then called a transfer press. These are particularly common in the automotive industry because of the high production quantities required and the complexity of many of the sheet metal parts used in modern vehicles, some requiring ten or more sequential operations.

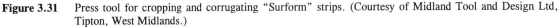

Figure 3.31 Press tool for cropping and corrugating "Surform" strips. (Courtesy of Midland Tool and Design Ltd, Tipton, West Midlands.)

Press tools are expensive to produce because of their complexity, the precision necessary and the costly hardened alloy steel from which they are usually made. However, their productivity is high, and they are therefore ideally suited to large-volume, continuous production.

3.7.2 Deep drawing

When items made from sheet metal are required, providing depth is not great relative to dimensions in the other two axes, it is usually not too difficult to cut and bend the material to the required shape. However, if depth is significant, the cold working process of deep drawing is used. What is considered to be a significant depth is a matter of personal interpretation within the industry, but most practitioners consider deep drawing to be required for circular components whose depth is greater than their diameter.

If the component is circular, spinning is an alternative to deep drawing, and this is described in section 3.7.3.1. While deep drawing is a much faster process, it can only be commercially considered if sufficient quantities are required to justify the high initial die costs.

3.7.2.1 The deep drawing process

A sheet metal blank previously cut to the correct size is forced, by a punch mounted in a press, into a suitably shaped steel die.

Theoretically, material thickness does not alter during drawing, but in practice some thinning does occur in the walls of the part produced, the degree of thinning depending upon the clearance between punch and die. This can be removed if necessary by a final light drawing operation termed ironing.

While deep drawing is by no means confined to the production of circular shaped items, it is easier to illustrate the basic principle involved by considering the drawing of a simple circular cup (Fig. 3.32).

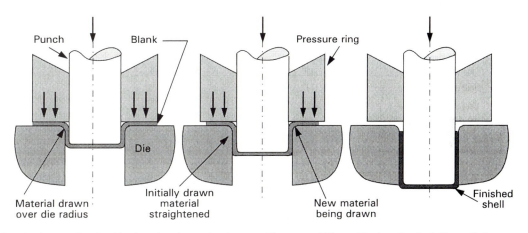

Figure 3.32   Basic steps involved in deep drawing a circular cup. (Courtesy of Platarg Engineering Ltd, Hanwell, London.)

The portion of the blank yet to be drawn down into the die is subject to a combination of heavy compressive and tensile forces as its peripheral size is reduced. The vertical wall of the pressing is under tensile stress as it is this part of the component that pulls the rest of the blank down into the die (Fig. 3.33).

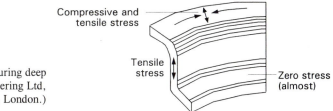

Figure 3.33   Simplified stress situation during deep drawing. (Courtesy of Platarg Engineering Ltd, Hanwell, London.)

To prevent the unbent blank from lifting and wrinkling during the drawing operation, a lubricated pressure ring presses down on top of the blank (Fig. 3.32). The pressure that this ring exerts must be sufficient to prevent the blank from wrinkling, but at the same time it must not be clamped so tightly that it prevents the blank from being drawn down into the die.

To establish the maximum reduction (maximum draw) possible in one operation, it is necessary to understand the metal's basic stress-strain characteristics within its plastic region, as this is where deep drawing occurs, i.e. beyond the elastic limit but below its fracture point – similar to the bending process described in section 3.7.1.2. Furthermore, as deep drawing is a cold working process, significant strain (work) hardening occurs, and this must also be taken into account. It is possible, using plasticity theory, to calculate the limits of draw possible for any given metal, but in practice it is very much a black art, with experience once more being a key factor. Furthermore, on high-speed transfer presses (up to 200 strokes/min), higher than expected draws are possible.

Drawing reduction ratio (DRR) is a term used to quantify the degree of reduction in one drawing step, and for circular components, is defined as:

$$\text{DRR} = \frac{(\text{Initial diameter} - \text{Final diameter})}{\text{Initial diameter}}$$

Exercise   Where have you come across this sort of relationship before? Do you think that the ratio of initial to final diameters would be an equally acceptable definition of DRR?

Industrial experience suggests that the first DRR, when working with fully annealed low-carbon steel, should be limited to around 35–40 per cent, as beyond this the risk of component fracture is too great. Commercial and technical judgement must be applied in deciding between the options of a number of progressively reducing drawing steps with no annealing required and fewer steps with interstage annealing.

To calculate the blank size needed to produce the required drawn component, it is necessary to equate the surface areas of the blank and the finished item, as it is assumed that material thickness stays substantially constant in deep drawing. Because this is not quite true in practice, slight modification to the calculated size is always necessary, based upon practical tests.

Such calculations are very simple, as is shown in the following example of a hemispherical bowl.

$$\text{Surface area of final bowl} = \text{surface area of hemispherical part} + \text{surface area of circular lip}$$

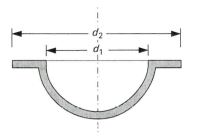

$$= \frac{\pi}{2}d_1^2 + \frac{\pi}{4}\left(d_2^2 - d_1^2\right)$$

Surface area of initial circular blank $= \frac{\pi}{4}D^2$

Equating surface areas, the initial blank size (D) will be at least $\sqrt{\left(d_1^2 + d_2^2\right)}$.

Figure 3.34 Profile of deep-drawn hemispherical bowl.

Assuming that the average tensile stress level during the deep drawing operation is midway between the material's yield and ultimate tensile stress, then the approximate punch force (F) required when drawing cylindrical parts is:

$$F = \pi dt \left(\frac{\sigma_{\text{ult}} + \sigma_{\text{yield}}}{2}\right)$$

where $\sigma_{\text{ult.}}$ is ultimate tensile stress, σ_{yield} is the stress at yield point, t is blank thickness and d is the diameter after drawing (Grainger undated). Practical experience has shown that the value of

F can be a considerable underestimate as it is strongly influenced by the degree of work hardening that exists in the blank before drawing.

One of the factors limiting the maximum DRR that may be used is the practical one that wear of rigid steel die-sets eventually occurs in the mechanical parts of the punch and die. This has the result that the punch does not press perfectly uniformly over the whole periphery of the blank, and this may cause localized high stress areas and consequently fracture of the part – usually in corner radii.

If a rigid and unyielding punch force could be replaced by one that is uniformly applied, it would then be possible safely to use a somewhat higher DRR, and consequently fewer draw stages would be required. This is possible thanks to the availability of a number of softer drawing/forming techniques that use such uniform pressure-transferring media as rubber and liquids (Fig. 3.35). These are referred to as flexible die-forming processes and, while they have the advantage that only one-half of a normal two-part die-set is required, the press equipment needed tends to be expensive. They are therefore usually only found in the aerospace and associated industries where small quantities of complex parts made in expensive metals are the norm.

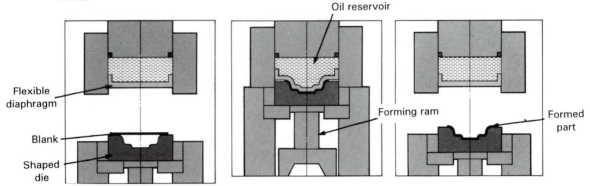

Figure 3.35   Hydraulic flexible forming process. (Courtesy of ABB Pressure Systems AB, Sweden.)

3.7.3 Spinning, shear forming and flow forming

This is a group of closely related processes designed to make symmetrical metal parts from circular metal blanks. They are all cold working processes, although the spinning of steel more than 30 mm thick is usually carried out as a hot working process to reduce the forming forces required.

As with all cold working, parts produced by any of these processes have a degree of work hardening and a refined granular microstructure in the direction of working.

3.7.3.1 Spinning

When the quantity of circular sheet metal components required does not justify the cost of expensive deep drawing dies, an answer may be to use the spinning process. This involves a circular blank of sheet metal being clamped against a steel or wooden former held in a heavy-duty centre lathe, copying lathe or CNC turning centre (§9.2.3). The former and blank are rotated, and a forming tool is then pressed against the revolving blank in a series of sweeping strokes until ultimately the blank is pressed into the shape of the former block (Fig. 3.36a). The forming tool is either hand held or gripped in the tool holder of a copying lathe, depending upon the size and material strength characteristics of the metal blank.

(a)

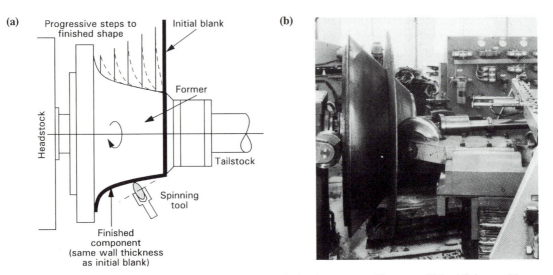

(b)

Figure 3.36 (a) Metal spinning principle. (b) Sheet metal spinning in progress. (Courtesy of Metal Spinners (Newcastle) Ltd.)

An important feature of spinning is that the thickness of the initial blank stays substantially the same throughout its transformation into the finished article.

A particularly important and quality-critical product range produced by spinning is dished ends used in the fabrication of pressure vessels (Ch. 5, Fig. 5.4).

3.7.3.2 Shear forming

Shear forming is a similar process to spinning, but with one very important difference. Here, the metal blank used is much thicker than the wall thickness of the end product, as during the forming process the metal is stretched beyond its elastic limit, resulting in its thickness being significantly reduced (Fig. 3.37a). This requires much greater forming forces than in spinning, so precluding any possibility of the forming tool roller used being hand held.

Since deformation of the metal only occurs at the point of contact between the shear forming roller and the blank, the remaining material remains stress free. This permits a much greater

(a)

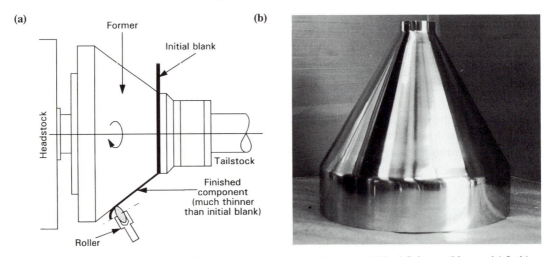

(b)

Figure 3.37 (a) Shear forming principle. (b) Typical shear formed item (Courtesy of Metal Spinners (Newcastle) Ltd.)

(a)

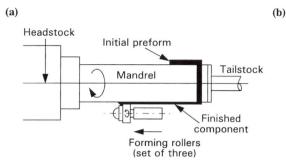

Headstock

Initial preform

Mandrel

Tailstock

Finished
component

Forming rollers
(set of three)

(b)

Figure 3.38 (a) Flow forming principle. (b) Typical flow formed components. (Courtesy of Metal Spinners (Newcastle) Ltd.)

degree of deformation to be achieved than is possible with other processes such as deep drawing, and usually results in the part being made in just one operation [Metal Spinners (Newcastle) 1994].

3.7.3.3 Flow forming

Flow forming is a variant of shear forming, and is specifically designed to produce cylindrical components. The process is the same as shear forming except that the initial blank is usually a preformed cylindrical cup (Fig. 3.38a). The process therefore differs from shear forming in that it involves elongating a thick-walled cup, as opposed to producing components from a flat cylindrical blank.

The preformed cup is produced by deep drawing or impact extruding if the quantities are sufficient. It can also be made by forging or turned on a lathe. Unlike shear forming a group of two or more forming rollers are normally required.

As the wall thickness of flow-formed parts can be varied at will over their profile (Fig. 3.38b), this process is ideally suited to the production of such items as thick-based saucepans. Certain high-pressure gas cylinder bodies are also produced by this process.

Extremely close internal tolerances can be maintained ($10\,\mu m$ or less in some cases), and surface finishes better than $5\,\mu m$ are regularly achieved.

3.8 Summary of the principal characteristics, advantages and disadvantages of metal deformation processes

Process	Characteristics	Advantages	Disadvantages
Rolling	The squeezing of cast metals into more useful and handleable shapes such as plate, billet, bar, sheet and structural sections	Improves and refines the metallurgical properties of the metal Provides shapes that are more convenient to the user and nearer to the final required form Excellent finish and accuracy with cold rolling	Poor surface finish and dimensional precision with hot rolling During cooling may lock in unwanted internal stresses which can cause unexpected distortions later Work hardening resulting from cold rolling creates a hard surface skin that may increase machining difficulties
Forging	The deforming of raw material into a shape as near as possible to the final required profile by imparting either impact (kinetic) or pressure energy	Greatly reduces material wastage and can be a much more economical option than casting or machining from the solid Impact strength and toughness are greatly enhanced, in part because of the beneficial grain flow pattern produced, which tends to follow the component's profile Forgings are usually easy to weld Allows one-piece components to be made which were previously fabricated assemblies	Manually operated systems are potentially hazardous to operators Only limited reshaping is possible when cold working, and there is significant danger of metal fracture if limits are exceeded Tooling is expensive so large-volume production is necessary to justify these costs. Also, forging presses are very expensive pieces of capital equipment
Extrusion	The pushing of metal through a hole in one end of a steel die, the profile of which determines the cross-sectional shape of the lengths of extruded material produced	Highly complex two-dimensional shapes are easily produced The die is readily changed so small batch sizes are economical Advantageous grain flow along the direction of extrusion is produced Dimensional accuracy and surface finish are usually good	Steel is difficult to extrude owing to its high compressive strength and special lubricants are needed Cold impact extrusion is severely limited in the degree of extrusion possible Three-dimensional shapes are not feasible
Cold drawing (thin rod, wire and tube drawing)	Similar principle to extrusion except that the metal is pulled through the hole in the die rather than being pushed through it	Small batches are feasible owing to the ease of changing the drawing dies Grain flow is in the direction of drawing Accuracy and finish are good Surface hardness and material toughness can be	Severe limitations on the degree of draw possible in one step, so interstage annealing may be required Because of the long lengths of product produced and the speeds that it is delivered from the die (up to 60 mph),

Process	Characteristics	Advantages	Disadvantages
Cold drawing (contd)		modified by varying process parameters	special mechanical handling facilities must be provided
Sheet metal bending	The bending of ductile sheet metal into final component shape; often also involves metal shearing both before and after bending	This is the only way to produce certain shapes such as car body panels Little or no further metal working is usually required – surface protection coatings can even be provided precoated on the material stock High production rates are possible when the process is mechanized	Dies are usually extremely expensive, and are only economical for large-quantity production, typical of that found in the automotive and "white goods" industries Complex three-dimensional shapes can be difficult to produce in one operation without the risk of metal failure Bends that are too tight cause serious thinning of the material, and there is then a risk of metal splitting in these areas
Deep drawing	A sheet metal bending process used when sheet metal parts are deep in relation to their dimensions in the other two axes	Accuracy and productivity are high when drawing operations are mechanized *Soft dies*, using elastomers or fluid as the drawing interface and top die, make small batch production possible, specially for parts with large surface areas. However, the special presses used for this are expensive	Because this is a cold working process there are severe limitations on the amount of draw possible in one step Interstage annealing is frequently a necessary expense Dies are very costly so large-scale production is essential
Spinning, shear forming and flow turning	Cold working processes used for the manufacture of precision circular parts, up to 4 m in diameter, from sheet metal/plate or preform	Refined microstructure is produced, giving enhanced physical properties With flow forming it is possible to vary component thickness as required In shear forming only the area being formed is under stress, so greater deformation is possible than with deep drawing Formers used in spinning are not expensive, and so provide an economic alternative to deep drawing if small batches are required	The metal blanks used in these processes must be capable of being rotated, i.e. not too large a diameter or mass For all but the most basic shapes special lathe-type forming machines are necessary, usually CNC controlled Considerable experience is needed if the part is not to fracture owing to excessive work hardening. Stress relief is an expensive additional operation, but is frequently essential if the final part is to be substantially free of internal stresses

Bibliography

British Steel Corporation 1992. *Technology and competitiveness – rolling and finishing*. Teesside: British Steel Corporation.

Grainger, J. A. Undated. *Brief description of the drawing of metal*. London: Dualform Division, Press and Shear Machinery.

Metal Spinners (Newcastle) 1994. Company internal report.

UK Aluminium Extruders Association 1991. *Aluminium extrusions – a technical design guide*. UK Aluminium Extruders Association.

Van Vlack, L. 1989. *Elements of materials science and engineering*, 6th edn. Reading, MA: Addison-Wesley.

Case study

You are required to devise a research test rig to measure the heat transfer characteristics of longitudinally finned aluminium tubing when carrying an exceptionally hot fluid at 400°C. To provide a source of circulating fluid at this temperature would be both expensive and potentially hazardous. It is therefore decided to simulate the hot fluid flowing through the finned tubing by resistance heating a thin-walled stainless-steel tube fitted into the bore of the finned tube (Fig. 3.39).

What problems would you anticipate in designing and manufacturing such an arrangement?

Note   The resistance heating principle is based upon the fact that all metals have a finite electrical resistance that resists the flow of electric current and in doing so generates heat. The amount of heat generated is readily controllable by regulating power to the resistance heating element – in this case a hollow steel tube.

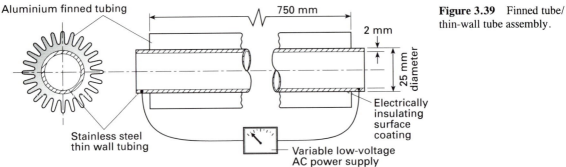

Figure 3.39   Finned tube/thin-wall tube assembly.

Solution

Because resistance heat must only be generated within the stainless-steel tube and not in the aluminium finned tube, it is essential that the two tubes are electrically insulated from one another. As any insulation layer will influence the fit achieved between the tubes, it is first necessary to establish precisely the tolerances of commercially available tubing.

Consider first stainless-steel tubing. This is produced by drawing (§3.6.2), and for the size of tube required here (25 mm outside diameter × 2 mm wall thickness) the commercial tolerance on both outside diameter and wall thickness is typically ±0.075 mm. Thus, the outside diameter can vary from 25.075 to 24.925 mm. Drawn tube can be externally ground to achieve a much narrower tolerance band, but because it is impractical also to grind such a long bore (it is 750 mm long!) grinding the outside diameter will result in an increase in wall thickness variation. This, in turn, will result in uneven resistance heat generated both circumferentially and along the length of the tube.

Aluminium finned tubing can be produced by either extrusion (§3.5) or drawing. Because extruded tubing is far less precise than that made by drawing, the commercial tolerance of 25 mm bore extruded tubing is typically ±0.3 mm, whereas drawn finned tube of the same size offers a bore tolerance of ±0.05 mm, albeit at a price premium of 250–300 per cent. (The reason for the higher cost and improved tolerance of drawn tubing is that drawn aluminium tubing is produced as a secondary operation performed on already extruded tube.) Here, there is little choice but to use the more expensive drawn tube, because if the extruded tube bore was at its minimum dimension of 24.7 mm it would be impossible to fit even the smallest size (24.925 mm) stainless-steel tube inside it. Fortunately, as this is a one-off research test rig, the additional cost is not critical in relation to the overall cost.

The extremes of fit when using both drawn stainless steel and aluminium tubing will therefore range from 0.125 mm diametral clearance to a diametral interference fit of 0.125 mm. Thus, if the insulation layer thickness is 0.065 mm, an interference fit between the two tubes will then always be achieved.

However, there would be a problem if the finned tubing supplied was of the smallest bore size available (24.95 mm) and the outside diameter of the stainless-steel tubing supplied was the maximum size (25.075 mm). Although this would result in a diametral interference of only 0.125 mm, application of an insulating layer of 0.065 mm would produce a total diametral interference of 0.25 mm. As the simplest method of tube assembly is to expand the aluminium tube by heating and then to slide it over the coated stainless-steel tube, the aluminium tube would have to be preheated to a temperature of at least 450°C to accommodate a diametral interference of 0.25 mm. Because aluminium melts at 660°C, such a high temperature is not recommended.

The simplest solution to this problem is carefully to measure the mean bore of the length of aluminium tubing to be used, and then apply a somewhat thicker insulating layer to the outside of the stainless-steel tube than is required. It can be centreless ground back to a diameter that will give a mean diametral interference of, say, 0.2 mm. This would require heating the aluminium to only 350°C before assembly.

Questions

3.1 Define hot and cold working as applied to the manufacture of engineering products. List the principal metallurgical properties that each imparts to steels.

3.2 In what way do metal deformation processes improve the finished product?

3.3 What are the main advantages of extrusion compared with the rolling of industrial sections?

3.4 Describe some of the defects that might occur during rolling of steel section. Why do they arise and what can be done to reduce their detrimental effects on the structural integrity of the rolled sections?

3.5 Discuss the main factors that determine whether a component should be cast or forged.

3.6 Discuss the principal advantages and limitations of the forging process.

3.7 Compare the metallurgical characteristics of forgings made using drop and press forging.

3.8 Explain why forging is often preferred to machining or fabrication. Refer specifically to a hexagonal-headed bolt and an open-ended spanner, using diagrams to amplify your explanation.

3.9 Briefly describe roll forging and explain its main differences from conventional hot rolling. Why is it particularly suitable for components made in a "chain form"?

3.10 A drop hammer weighing 500 kg has a free fall of 750 mm. Calculate:
   (a) the energy of the hammer blow at the instant of impact
   (b) the velocity at the instant of impact
   (c) the average force exerted by the hammer if it deforms the forging 2.5 mm in one blow
   (d) the final area of the forging if it has an average yield strength of $80 \, MN/m^2$ at forging temperature.

3.11 What advantages does impact extrusion have compared with deep drawing when producing cup-shaped objects? Why is it not used for the manufacture of all such items? How would a high rate of production be maintained?

3.12 Compare the advantages and disadvantages of direct extrusion and indirect extrusion.

3.13 There is a problem common to the manufacture of both thin wire and small-bore tubing when being drawn down. What is this difficulty and how has it been solved?

3.14 The amount of reduction per draw possible when cold drawing wire from a significantly larger diameter is severely limited by the changes that take place in the metallurgical properties of the wire. What are these principal changes and what can be done to increase the degree of draw that is possible at each pass?

3.15 Using simple sketches, illustrate the major stresses to which a sheet metal blank is subjected when being deep drawn into a circular cup whose depth is significantly greater than its diameter.

3.16 What advantages does flexible die forming have over forming using conventional steel dies?

3.17 What is meant by the terms drawing reduction ratio and multistage drawing? When multistage drawing is found to be necessary, what other process must also be carried out and why?

3.18 Derive an expression for the overall drawing reduction ratio (T) in a multiple drawing operation in terms of (n) individual drawing reduction ratios (R). Base your derivation upon a circular cup type of component as illustrated in Figure 3.32, and take

$$R = [(d_0 - d)d_0]$$

where d_0 and d are the blank and punch diameters respectively.

3.19 Why, in a blanking or piercing operation with a press, might the punch or the die have shear applied to it? Explain clearly the meaning of shear in this context.

3.20 What is meant by the term springback, and how is its effect catered for in forming operations?

3.21 In blanking operations, sketch and describe examples of where component layout can yield the maximum number of parts with the minimum of scrap.

3.22 A hole 100 mm in diameter is to be punched in a mild steel plate 6 mm thick, the ultimate shear stress of the material being $550 \, MN/m^2$. Piercing is complete at 40 per cent penetration, the radial clearance on the tooling being 8 per cent of the material thickness. The capacity of the available press is 200 kN. Determine:
   (a) suitable punch and die diameters
   (b) the shear angle necessary if the above press is to be used.
   Discuss the validity of any assumptions made in your calculations.

3.23 Discuss the application of metal spinning and compare its advantages with those of other metal forming processes that could be used to produce a hemispherical telecommunications dish.

3.24 What are the principal differences between:

(a) spinning and shear forming

(b) shear forming and flow forming?

List examples of where each process would be an appropriate method of manufacture.

Chapter Four
Processing of plastics and elastomers

After reading this chapter you should understand:
- (a) what a plastic is and that there are only two basic types
- (b) how plastics are turned into useful products
- (c) the principal limitations of plastics compared with metals
- (d) why there is increasing interest in laminates and composites
- (e) the most important design guidelines when designing plastic components
- (f) what an elastomer is and how it differs from all other engineering materials
- (g) why metallic inserts are frequently incorporated within plastic and elastomeric parts
- (h) the properties of natural and synthetic rubbers and why they are particularly suitable for many engineering applications.

4.1 Plastics – the two families

Plastics may be divided into two basic types: (a) thermoplastics and (b) thermosetting plastics, also called thermosets. To appreciate the many ways in which plastics may be used in engineering, it is not essential for the manufacturing engineer to understand in detail the underlying chemistry (Smith 1990). Nevertheless, it is important that one fundamental difference between these two families of plastics be understood as it significantly influences the way in which each can be processed.

4.1.1 Thermoplastics

Heating thermoplastics softens (plasticizes) them, and if further heat is applied they eventually melt. Upon removal of the heat source, as they cool down, they regain their original hardness. Thermoplastics can be replasticized/remelted any number of times by successive applications of heat without risk of degradation of the material. Chemically the material is unchanged, irrespective of its temperature, up to the point where it is heated to such an extent that it is destroyed.

Typical examples of familiar thermoplastics are: acrylics (Perspex and Plexiglas), PVC, polystyrene, polythene, PTFE, nylon, polypropylene and fluorocarbons.

4.1.2 Thermosetting plastics (thermosets)

Unlike thermoplastics, thermosets undergo an irreversible chemical change when heated, so that the material, after heating, is no longer chemically the same as it was initially. This permanent chemical change, called *polymerization* (Smith 1990), causes the material, after initial softening/melting, to harden when still hot, and this has a major influence on the processing methods that can be employed compared with processing thermoplastics. Reheated thermosets never resoften and retain their hardness up to the point where they are destroyed by charring and finally burning, i.e. they can be considered to be permanently set.

One major disadvantage with thermosets is that, unlike thermoplastics, they cannot be reused because, as the initial chemical composition is permanently changed by polymerization, further working of the material is impossible.

Typical examples of common thermosets are: Bakelite, polyesters, melamine and epoxy resins such as Araldite.

Exercise   Over the next week make a list of the day-to-day items that you come across that you think are made from some form of plastic. Carefully file this list until you have completed this chapter. You will then be asked to retrieve it and use it as the basis for further work (Question 4.1).

4.2 Plastic forming processes

Because thermoplastics when heated, and thermosets when in their initial pre-polymerized form, are all physically highly flexible, they lend themselves well to being moulded into a wide variety of shapes. In some respects this is similar to the moulding of liquid metals (Ch. 2), except that plastics are not liquid when moulded.

There are a wide variety of processes available for making components from plastics (Pearson 1985), but only the most commonly used ones are described in this chapter. Particular attention should be paid to how the fundamental chemical difference between the two plastic families influences the ways in which they can be processed.

4.2.1 Compression moulding – thermosets only

The required component is moulded in a steel die-set using a design approach similar to that used in non-ferrous die casting (§2.5), i.e. the die must be highly polished to provide a blemish-free surface finish on the components produced and be readily openable to eject the finished component (usually by ejector pins) without damage to either die or part.

The principle involved is as follows. A precise quantity of plastic raw material (the charge), in either a granular or pelletized preform condition, is introduced into the cavity of a preheated die (Fig. 4.1a). A die plunger is then lowered onto the charge which, because of the heating effect of the hot die, is plasticized sufficiently to be moulded into the required shape. Pressure and die heating (120–220°C, depending upon the thermoset being used) is maintained until full polymerization (curing) has been achieved, whereupon the die is opened and the finished component ejected. The open die is then ready to receive its next charge.

Polymerization can take anything from 30 seconds to several minutes, depending upon the

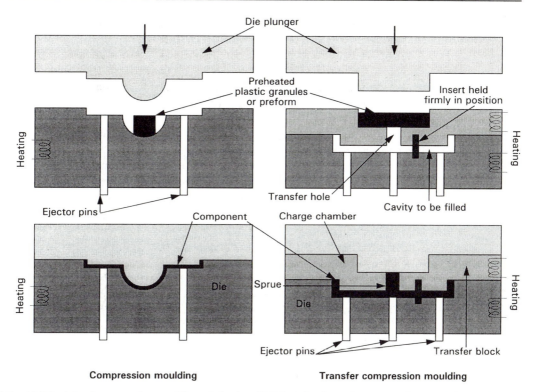

Die plunger

Preheated plastic granules or preform

Insert held firmly in position

Heating

Heating

Heating

Transfer hole

Cavity to be filled

Ejector pins

Component

Charge chamber

Heating

Die

Sprue

Die

Heating

Ejector pins

Transfer block

Compression moulding

Transfer compression moulding

Figure 4.1   (a) Principle of compression moulding of plastics. (b) Principle of transfer compression moulding of plastics.

complexity of the part and whether or not the charge is preheated. Preheating can halve production time as well as reduce die wear, and this is one reason why the preheated preform format of charge is so popular.

Note that the component is still hot when ejected from the die, but it is able to retain its shape because of the unique chemical change characteristic of thermosets (§4.1.2).

Although compression moulding could be used to mould thermoplastic materials, each component and the die would have to be cooled sufficiently to ensure that the finished part retained its shape when it was ejected from the die. The die would then have to be reheated before the next component could be made. Such cyclic heating and cooling of the die-set is uneconomic, and this is why compression and transfer compression moulding (§4.2.2) tend to be used only for moulding thermosets.

4.2.2 Transfer compression moulding – thermosets only

Transfer moulding is basically a slightly more sophisticated form of compression moulding, enabling greater positioning accuracy of moulded-in inserts (§4.3.1), as well as ensuring a more even moulding pressure over the whole surface of the component. This, in turn, results in less die wear, more uniform polymerization and the ability to produce parts that have thinner sections than would be possible with straight compression moulding.

Die design is similar to that used in compression moulding except that the plunger profile no longer forms part of the die cavity that determines component shape (Fig. 4.1b), but conforms to the recess in a heated charge chamber introduced between die and die plunger. Before each closure of the heated die-set, following ejection of the previous part, any inserts that must be moulded into the next part are placed in the bottom die. The die is then closed, and a charge introduced into the charge chamber. When it has plasticized sufficiently it is injected through a transfer hole into the die cavity. When polymerization has taken place the die is opened and the component ejected and allowed to cool. Finally, the sprue that is formed in transfer moulding (but not in compression moulding) must be removed by hand.

4.2.3 Injection moulding – mainly of thermoplastics

This process differs from compression and transfer moulding in that it provides the moulding process with an almost continuous supply of heated plastic. It also provides for rapid cooling of the finished part.

When working with thermoplastics, unlike thermosets, it is essential to cool the finished part as quickly as possible to a sufficiently low temperature to ensure that it will retain its shape when ejected from the mould. This is the only way to achieve short mould cycle times.

The process involves gravity feeding raw material from a hopper into a pressure chamber ahead of a plunger (Fig. 4.2a). As the plunger advances, granular plastic is compressed and then forced through a heating cylinder to raise its temperature sufficiently to bring it to a

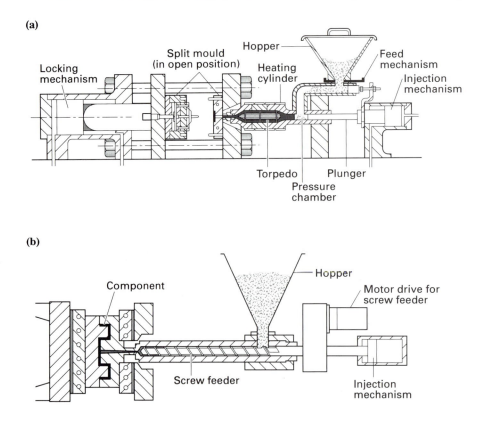

Figure 4.2 (a) Ram feed injection moulding. (b) Screw feed injection moulding. (Courtesy of Tripwall Ltd, East Tilbury.)

semiliquid/plasticized state. Uniformity of heating is essential and is greatly assisted by the inclusion of a torpedo-shaped object mounted in the centre of the heating cylinder, as this generates an annular section of material to be heated rather than one of solid circular cross-section, which would be much more difficult to heat evenly over its whole thickness. The plasticized material is then squeezed through an ejector nozzle at great pressure into the die cavity to form the required component. The die section is heavily water cooled, making the injected plastic freeze almost immediately the die cavity is filled.

Quite large thermoplastic components, such as refrigerator door linings, may be made by injection moulding.

As with high-pressure injection moulding (die casting) of metals (§2.5.3), die costs are high so large production runs are essential if component cost is to be competitive. However, this process is easily automated, and cycle times of just a few seconds are common. This, in part, explains why the majority of plastic items are produced by injection moulding.

A highly effective modification to the above feeder system is frequently employed. It involves replacing the feed plunger with a motor-driven screw plasticizer (Fig. 4.2b). This type of drive is popular because, by altering the pitch of the screw and its core diameter at certain points along its length, the thermoplastic can be subjected to a controlled shearing action that both expels trapped air and supplements the heating process provided by the external heaters. It therefore helps to ensure that the thermoplastic fed through the injector nozzle is maintained at a constant and uniform temperature/viscosity. Using the correct screw contour is essential as different contours are required to suit the unique thermal and extrusion characteristics of different thermoplastics. Fortunately, changing an injection moulder's screw is not too difficult a task.

4.2.4 Extrusion – mainly of thermoplastics

Screw-type extrusion machines, similar to those used in injection moulding, are used as the source of supply of a continuous stream of plasticized plastic. The basic principle of the process is the same as that used in the direct extrusion of metals (§3.5.1), namely the feed material is forced through a die whose orifice is the same shape as the required extrusion (Fig. 4.3). Lengths of a wide range of plastic sections are produced, ranging from curtain rails and PVC water piping to the coating of wire.

Thin plastic sheet and film can also be made by extrusion. This is achieved by extruding a circular shape, inflating it with compressed air until the plastic has cooled, and then removing the air supply so that the tube deflates. It is then rolled flat and its sides are either slit to form two thicknesses of sheet or chopped off in lengths for use as feedstock in blow moulding (§4.2.6). However, most thermoplastic sheeting is made by the calendering process (§4.2.5).

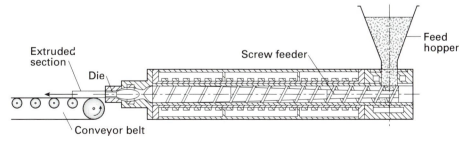

Figure 4.3 Extrusion of plastic sections. (Courtesy of Tripwall Ltd, East Tilbury.)

4.2.5 Calendering – of thermoplastics

This process is used to produce continuous lengths of thermoplastic sheet. The calendering unit (Fig. 4.4b) comprises a series of heated rollers through which is passed a continuous stream of heated plastic fed from a feed hopper placed above the first of the rolls. As the plastic passes through the rolls it is squeezed into a continuous sheet and emerges to be either wound onto storage rolls or cut into convenient lengths by flying shears (a knife that moves at the same velocity as the moving sheet so that the cut produced is at right angles to the material's feed direction).

(a)

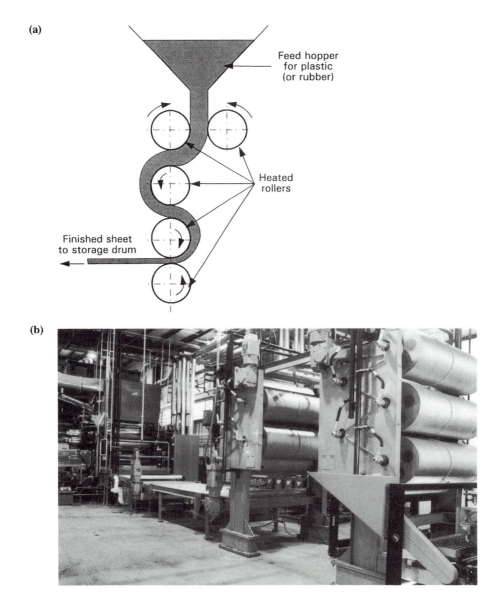

(b)

Figure 4.4   (a) Principle of calendering. (b) Plastic sheet calendering line. (Courtesy of Farrel Ltd, Rochdale.)

4.2.6 Blow moulding – of thermoplastics

The blow moulding process involves four operations:
1. A two-part die is opened and a heated length of thermoplastic tubing (known as a parison) is either drawn onto or extruded over an air pipe placed between the two halves of the open die (Fig. 4.5a).
2. The die halves are then closed and, in so doing, nips shut the open end of the parison.
3. Air is next blown into the parison, and this expands the plastic tube out to the walls of the die (Fig. 4.5b), which are usually water cooled to cool the plastic down as quickly as possible, and so minimize cycle time.
4. Finally, the die is opened to allow the component to be ejected (Fig. 4.5c).

This forming process is easily mechanized, and is often integrated with an extrusion machine that is producing a continuous supply of tubing. It is particularly useful for the high-speed production of bottle-shaped components, from everyday small flexible domestic bottles to large-capacity drums.

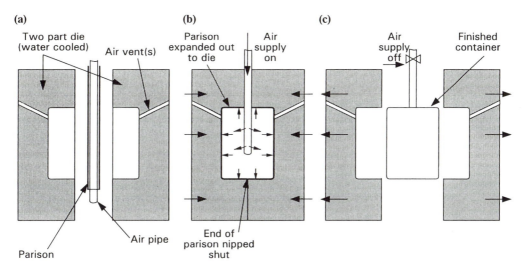

Figure 4.5 Principle of blow moulding plastics.

4.2.7 Vacuum forming – of thermoplastics

In one respect vacuum forming is very similar to blow moulding in that it involves the application of a differential pressure across the plastic raw material (Fig. 4.6). A preheated sheet of thermoplastic is placed in a one-piece die, a sealing plate is attached, and the atmospheric air within the mould is sucked out by means of a vacuum pump. This causes the plastic to collapse onto the mould and conform to its profile. When the plastic cools the pump is switched off, the sealing plate is removed and the finished moulding lifted out of the die.

With vacuum forming clearly the maximum differential force possible is only 1 atmosphere, and this limits the thickness of sheet stock that can be used. Even so, the total forming force can be significant when the surface area of sheet used is large.

Products made via this process range from bath and shower cabinet mouldings to preformed product packaging materials.

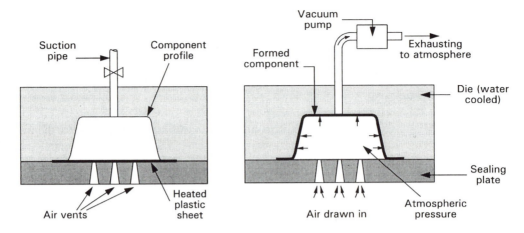

Figure 4.6   Principle of vacuum forming plastic sheet.

4.2.8 Laminates and composites

One severe limitation of all plastics is that they are structurally very weak compared with most metals. This is a pity as they have many extremely attractive attributes: they are of low density, low cost and have an attractive appearance; component manufacture is easy; minimal post-production machining is required; and, in the case of thermoplastics, there is virtually no waste material. Combining plastics with other materials(s) to produce new composite materials whose properties are superior to the original plastic, i.e. designer materials, is an exciting option that is now available to today's designers, although simple reinforced composites have

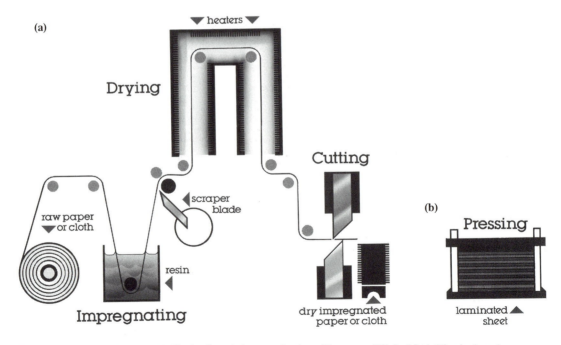

Figure 4.7   (a) Prepreg manufacture. (b) Flat laminated sheet production. (Courtesy of Tufnol Ltd, Birmingham.)

been available for many years in the form of laminated sheet, tube and rod. These are made using various non-metallic materials such as different grades of paper, woven cloth, fibreglass, etc., which are impregnated with thermosetting resin (prepreg as it is commonly known) (Fig. 4.7a), and then built up in layers to the thickness and shape required (Fig. 4.7b). Finally, the layers are compressed in a heated oven to complete polymerization and to form a consistent texture throughout the whole thickness of the laminate.

An example of this process frequently carried out by hand is in the production of a sailing boat hull: fibreglass woven matting is laminated in thermosetting plastic, which is manually applied by brush onto a suitable mould former.

Safety warning   Many plastics emit toxic fumes. It is therefore essential, when working with such materials, to take adequate precautions to avoid health problems. These may include wearing a face mask and ensuring that the work area is well ventilated.

Sheet, tube and rod laminates are all readily available and can be used to manufacture machined engineering components such as gear wheels for non-critical applications. Because the range of modern composites is so great and growing all the time, it is claimed that, by the year 2000, 30 per cent of all structural materials will be composites rather than traditional metals such as steel and aluminium (Banks 1994). It is also suggested that by the turn of the century 50 per cent of modern combat aircraft will be made from composites. It is therefore vital that today's design engineers know how to design components for manufacture in plastic and composite materials.

Note   Only a few years ago the fibreglass body of the average Formula 1 grand prix racing car would only last for one race and associated practice sessions. With the introduction of carbon fibre as a reinforcing material, skilfully applied in the areas of greatest dynamic stress, such car bodies now last a whole grand prix season.

4.3 Design considerations when designing plastic components

A common mistake made by inexperienced designers is to design components to be manufactured in plastic in the same way as they would if metal was to be used. While they will certainly allow for the big differences in stress-strain characteristics, ultimate shear stress and other relevant physical parameters between metals and plastics, frequently no account is taken of the special demands of the actual manufacturing processes involved. This results not only in avoidable problems on the shop floor, but also in failure to achieve fully the considerable benefits of using plastics correctly.

Notwithstanding this, there are certain general design rules that designers familiar with designing cast-metal components (§2.7) should equally well apply when designing plastic parts. Examples of these are:

  – Never call for sharp corners as they are one of engineering's major stress-raising features which, in turn, leads to the formation of stress cracking. The largest sensible radii should therefore always be used.

91

- Maintain as uniform a wall thickness as possible, as uneven cooling is a great problem with plastics owing to their low thermal conductivity.
- Avoid large sections being fed from smaller ones.
- Provide draft angles to aid component ejection from the die – usually 3° is sufficient.
- Allow for contraction of the component material; typically most plastics have a coefficient of expansion at least double that of ferrous metals (i.e. 0.000024 per °C or more).
- Design the component to avoid areas that could cause gas entrapment during injection of the plastic.
- Ensure that the component can be easily ejected from the die-set.
- Avoid undercuts and re-entrant shapes to keep die design as simple as possible. Die costs are high anyway, so there is no point in making them even higher.
- Ensure that minimum safe wall thicknesses are maintained, typically 1.5–2 mm.
- Try to avoid the need for any machining if at all possible. There are two important reasons for this. The first is that thermoplastics in particular are notoriously difficult to cut owing to their tendency to soften at temperatures that are all too easily reached at the cutting tool tip. (The special problems of machining plastics are considered in more detail in Chapter 8, §8.10.) The second reason for minimizing machining is not quite so obvious. Most plastic parts when moulded have a thin, shiny, resinous film all over their outer surface. Post-moulding machining, if carried out, removes that shiny surface and exposes a much less attractive matt one. Although this does not matter in areas such as screw holes, the removal of sprues, if they are not positioned carefully, will leave unsightly areas in full view. As an attractive appearance is usually an important factor with plastic components, special care must be taken in their design to avoid such problems.

Exercise   Take the removable cap of a cheap, throwaway, plastic ballpoint pen and examine its surface finish. Look carefully for evidence of flash and sprue removal. (If you can find none the die designer has done a good job!) Now take a penknife or razor blade and carefully scrape away a little of the shiny surface. What you then see is the matt finish referred to above.

- Core holes wherever possible – their positional accuracy and surface finish will be much better than that achievable by subsequent machining.
- As with sheet metal, large flat plastic surfaces require stiffening, and the use of ribbing helps to overcome this problem. Indeed, design appearance can be enhanced by clever use of rib features.

Exercise   Examine the shape of car bonnets and floor pans. Can you see where ribbing has been used to advantage both structurally and visually?

- Large, featureless expanses of plastic have a tendency to show moulding flow lines, but these can be also be camouflaged by the use of ribbing.
- Countersink all holes slightly, especially if they are to be threaded or contain metal inserts (§4.3.1). This is always a better option than cutting threads in the raw plastic.
- Ensure that the material thickness around any inserts used is sufficient to withstand the loads applied to the insert.

Some composites may have, by design or by nature, physical properties that are highly directional, and this must be taken into account at the component design stage. While offering good strength and stiffness to mass ratios, fibre strength in the direction of the reinforcement can be exceptionally high, but in all other directions it may be only that of the basic plastic. An extreme example of this is a carbon fibre composite in which the fibres are laid in certain directional planes only. Carbon fibre has a modulus of elasticity (E) three times that of carbon steel, but the thermosetting resin in which it is usually encased has an E value of only approximately 0.004 that of steel. This is why chopped fibre is used for random directional reinforcement of many composites; alternatively, alternating layers of directional reinforcement can be used to minimize variations in directional strength.

4.3.1 Metal inserts in plastic components

It has already been stated that machining plastics is difficult (§4.3) and should be avoided if at all possible. This implies that as many as possible of the required features should be moulded. Unfortunately, one of the most common features required is threads, which by their nature are extremely difficult to mould accurately and are just as difficult to machine. Even if it were possible to produce good accurate threads in plastic, their strength would be very low and they would be likely to strip when tightening metal screws in them using even modest force. A simple solution to these difficulties is to mould into the part threaded metal inserts, usually made from brass or steel. Both male and female threaded features can be provided, and their accurate location can also be guaranteed (§4.2.2).

Metal inserts are not restricted to threaded items, and the same principle can be used, for example, to provide the part with precision location dowels (lengths of precision-ground circular rod) to ensure accurate positioning of two or more mating parts.

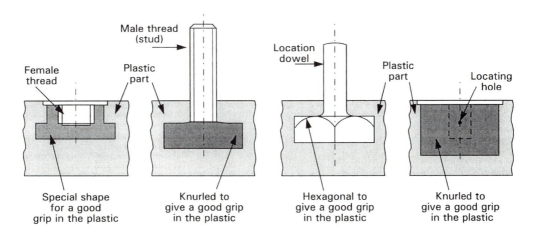

Figure 4.8 Typical forms of metal insert used in plastic moulding.

4.4 Elastomers

An elastomer is the name given to a material that has the ability to deform elastically under load to a much greater degree than metals, and then almost to recover its original form when the load is removed. The best-known material in this category is rubber, the basis of which is natural latex – a white milky substance obtained from rubber trees.

Exercise   Pick a dandelion and squeeze its stem. The milky substance obtained is also a latex, although not quite the right sort for producing good-quality natural rubber. Even if it was of acceptable quality, harvesting would be much more laborious and difficult than the tapping of rubber trees every 2–3 days.

The production of synthetic (artificial) elastomers that mimic the chemical composition of natural rubber has long been a major activity in the petrochemical industry, and the bulk of all elastomers used today are synthetic. One of the greatest benefits of synthetic elastomers is that they can be easily modified chemically by the addition of other materials during their manufacture. This helps to obtain the required properties.

As with plastics, it is not necessary for the mechanical design engineer to possess a detailed knowledge of the organic chemistry involved in turning latex into useful forms of rubber or to understand how modern synthetic elastomers are made. It is only necessary to appreciate how elastomers can be most effectively employed as engineering materials.

Exercise   Retrieve the list of items made from plastic that you were asked to produce in the exercise in section 4.1.2. Add to this list as many applications of elastomers that you can think of – the modern motor car is a good place to start!

4.4.1 Elastomer component production

Both rubbers and synthetic elastomers readily lend themselves to being moulded and extruded into the required shapes. With few exceptions most elastomers can be likened to thermosetting plastics in that they undergo an irreversible chemical change when processed. It is therefore not surprising that many of the methods used to produce thermosetting plastic components are equally suitable for the manufacture of elastomeric parts. Processes used for working thermoplastics are also used where appropriate.

Compression moulding (§4.2.1) and injection moulding (§4.2.3) are used to make parts of complex shape such as engine oil seals, and extrusion (§4.2.4) is employed to make long lengths of complex cross-section typical of that used for car door weather seals. Calendering (§4.2.5) is used to produce both natural and synthetic rubber sheeting.

While designers should be aware of the types of elastomer best suited to the most common engineering applications (Freakley & Payne 1978), there is no substitute for seeking the advice of leading elastomer producers for a given intended application. Factors such as high coefficient of expansion (for most elastomers this is ten times that of steel) are then less likely to be overlooked during component and die design.

Figure 4.9   Typical range of elastomeric items. (Courtesy of GMT Rubber Ltd, Leeds.)

4.4.2 The most commonly used elastomers

4.4.2.1 Natural rubber (NR)

Despite the wide range of synthetic elastomers available, NR is still the best choice for certain applications. It is particularly suitable for spring/suspension and anti-vibration mounting duties providing the environment does not include oil or solvents or is not subject to extremes of temperature. A large choice of hardnesses is available, and it is particularly easy to form to the required shape (e.g. windscreen wiper blades). Road tyre production is the largest single use of NR, partly because of its high coefficient of friction.

Note   Unlike normal tyres the tyres of a racing car are sticky to the touch. This is because the particular blend of NR used contains special oil and other additives designed to maximize the coefficient of friction at racing temperature (90–100°C at the tyre's surface; J. Jentgen, works director of Goodyear Tyre Co., personal communication). However, the penalty is extremely high wear rates.

4.4.2.2 Synthetic rubbers (SRs)

There is a vast and ever-growing range of synthetic rubbers, the eight referred to below being only examples of some of those most commonly used in engineering. They should, however, provide a good indication of the class of elastomer most likely to be suitable for many day-to-day applications.

– Styrene-butadiene rubbers (SBRs) are the most common group of synthetic rubbers as they are readily compounded with various fillers to offer a wide choice of end properties. Because they are cheaper than NR and are easily mouldable they are a popular choice for high-volume, low-cost items, but they do not withstand oily environments, although they do offer better water resistance than NR. They are also used as a minor ingredient in tyre manufacture.
– Polybutadiene rubber (BR), unlike NR, offers good low-temperature flexibility and high abrasion resistance, which makes it particularly suitable for the soles and heels of boots

and shoes. It is also another minor ingredient in tyre manufacture.

- Nitrile-butadiene rubber (NBR) has particularly good oil resistance, which makes it an appropriate choice for "O" rings and oil seals.
- Isobutyl rubber (IBR) is particularly weather, heat and chemical resistant. It is also highly impermeable to gases, making it ideally suited to the manufacture of tyre inner tubes.
- Neoprenes, although having poorer physical properties than NR, offer superior chemical, oil and elevated temperature (100–120°C) resistance. They are commonly used in the manufacture of belting, oil seals, garage petrol pump hoses and oil-resistant cable insulation.
- Silicone rubbers have exceptional low- and high-temperature resistance (–90 to +250°C), but this is only achieved at the expense of having many physical properties that are inferior to other elastomers. Silicone rubber is very expensive, and its use is therefore mainly confined to difficult high- or low-temperature sealing applications, typical of those found in the aerospace industry. Its other principal use is for the production of moulds in which a high degree of precision and detail is required, particularly in the art world.
- Polyurethane rubbers (PURS). Because of their good abrasion resistance and high tensile and tear resistances, they are particularly suitable for flexible die-forming material in deep drawing operations (§3.7.2.1). They are also available in a viscous liquid form, making them ideal for use as a casting medium to produce a wide range of parts ranging from trolley wheels to car bumpers and rubbing strips.
- Fluorocarbons (e.g. Viton) are elastomers that are extremely resistant to petrochemical attack even at elevated temperatures, which makes them especially suitable for use in fuel lines and associated sealing elements. They are also used as a covering for printing rolls.

4.5 Summary of the principal characteristics, advantages and disadvantages of plastics processes

Process	Characteristics	Advantages	Disadvantages
Compression, transfer and injection moulding	The forcing of a plastic into a cavity having the same shape as the required component	Raw material is usually much cheaper than metallic alternatives Colour and other physical properties not difficult to manipulate Injection moulding, in particular, is readily automated, giving component cycle times measured in seconds Inserts are easily incorporated into the moulding Thin-wall sections (typically 1.5 mm) are consistently achievable The precision possible often eliminates the need for any machining Little material is wasted, but what there is tends to be the non-recyclable thermoset type	Moulds are expensive as they must have a mirror-like finish if components are to have a blemish-free surface Only thermoplastics can be injection moulded, thermosets being processed by the slower compression/transfer moulding methods Flash removal is necessary and must be sited so as to not mar component appearance

Process	Characteristics	Advantages	Disadvantages
Extrusion	The squeezing of plastic through a hole in a die to produce continuous lengths of product of the desired cross-section	Highly complex shapes are easily produced in any required length Excellent finish and accuracy are possible Ideal for producing plastic extruded tubing suitable for feedstock for blow moulding machines	Limited mainly to thermoplastics owing to the higher cost of thermosets and the length of curing time required
Calendering	The production of continuous lengths of plastic sheet	Simple process suitable for both plastics and elastomers	The equipment is generally large, heavy and can be costly to both buy and maintain
Blow moulding	A process for high-speed production of plastic bottle-shaped items, using compressed air as the forming medium	Production rate is high, and as thermoplastics are the feedstock unit costs are normally low No limit to the forming pressure that can be used (up to 6 bar is normal), so significant detail can be reproduced, including screw threads No finishing operations are normally necessary	Requires a constant feed of tubular material, and this usually means an extrusion unit is necessary either at the input feed end of the moulding machine or built into the machine itself Moulds can be expensive, but the quantities produced usually more than compensate
Vacuum forming	Similar process to blow moulding except that a vacuum is used to press the plastic against the mould rather than air pressure. Parts so produced are usually made from flat sheet rather than from tube	Ideally suited to the production of large surface area parts from relatively thin sheet, such as bath and shower cabinet mouldings Also used for the manufacture of preshaped packaging materials	Only a maximum of 1 atmosphere of moulding pressure is possible. This limits the material thicknesses that can be successfully moulded

Laminates and composites	The manufacture of parts using plastics that have other material(s) embedded within them to enhance their physical properties	A wide range of composites are available
		   Properties can be modified to optimize the composite's suitability for any chosen application
		   Composites are increasingly being developed that can safely replace metals because of cost, life, and mass to strength ratio advantages
			Some composites can be expensive
			   Special care is required when working with composites having directional characteristics, as the variations in physical properties with respect to direction can be enormous
			   Laminates can produce toxic fumes, so when applied manually health precautions are necessary
			   Many plastic composites have a great affinity for moisture, and have coefficients of expansion more than twice that of steel

Bibliography

Banks, W. M. 1994. A future for advanced composites: dream or reality?, *Professional Engineering* **7**(9), 14–15

Freakley, P. K. & A. R. Payne 1978. *Theory and practice of engineering with rubber*. London: Applied Science.

Pearson, J. R. A. 1985. *Mechanics of polymer processing*. London: Elsevier Applied Science.

Smith, W. F. 1990. *Principles of materials science and engineering*, international edition. New York: McGraw-Hill.

Tufnol 1985. *Gear design in tufnol laminates*. Birmingham: Tufnol.

Case study

A manufacturer of Roots-type air blowers (Fig. 4.10) sells much of its output for use in pneumatic conveying systems (i.e. transporting powdered or small-particle substances along a pipeline using a large volume of low-pressure compressed air). A proposed new application is to unload pneumatically road tankers that deliver flour to large bakery silos.

The current cast iron blowers were deemed to be too heavy for fitting to the tankers, and it was suggested that aluminium castings be used instead. This proposal proved successful from the weight reduction point of view, but presented noise difficulties not previously encountered with cast iron body and gearbox castings.

What proposals would you make to reduce the unacceptable pneumatic and gear noises so that the new lightweight blowers complied with noise control regulations?

Solution

The pneumatic noise emanating from the aluminium body casting was reduced to an acceptable level by lagging the outside surface of the casting with a suitable sound-deadening material. Unfortunately, similar treatment to quieten the aluminium gearbox was not possible because natural convective cooling was necessary to ensure that the gearbox oil did not overheat. It was therefore suggested that the gears be made from a plastic composite material (§4.2.8), rather than from hardened steel. However, this was difficult to achieve in practice as one of the existing steel gears was specially designed to allow for precise timing of the two blower impellers, such that they ran as close together as possible but without actually touching one another when under load.

It was felt that a major reduction in gear noise could still be achieved if only one of the gears was made from a composite while the other (steel) gear could still be used for impeller timing purposes and so avoid changing the proven method of timing adjustment used with pairs of steel gears.

Two big differences that directly affect gear design when using plastic composites instead of steel are the former's much lower safe working stress and its much greater coefficient of linear expansion. The lower working stress of composites can be compensated for by increasing gear face width, but to quantify this the composite manufacturer's advice and guidance was obtained (Tufnol 1985). This resulted in a gear almost twice the width of a comparable steel gear, but, more importantly, the lamination format of the composite from which the gears were made had also to be taken into consideration (Fig. 4.11).

Figure 4.10   Roots-type air blower. (Courtesy of Dresser Roots, Huddersfield.)

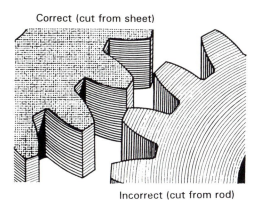

Figure 4.11   Lamination format in composite gears.
(Courtesy of Tufnol Ltd, Birmingham.)

A coefficient of linear expansion approximately double that of steel was a problem as the gears had to be provided with sufficient backlash (free movement between two mating gears) when cold to take account of the extra radial expansion of the composite gear compared with a steel one. If this had been ignored, the gears would have jammed in mesh when at working temperature, causing excessive wear. Indeed, the precise amount of extra backlash provided was critical: too much could have resulted in the two impellers hitting one another until the timing gears had warmed up, by which time the blower would have suffered serious damage.

Questions

4.1 Retrieve the list of items made from plastic and elastomers that you were asked to produce in the exercises in sections 4.1.2 and 4.4. In the light of what you have learned by studying this chapter, suggest a suitable method of production for each of your plastic and elastomer applications; do not omit to take account of the likely quantities produced and how this might affect the choice of process. Also suggest a suitable material for each item on your list.

4.2 What is a plastic and how does it differ from an elastomer?

4.3 What is the greatest single limitation to the use of both plastics and elastomers in engineering?

4.4 Explain the important differences between the two basic types of plastic. How are methods of component manufacture influenced by these differences?

4.5 What are the main advantages and disadvantages of transfer moulding of plastics and rubbers compared with straight compression moulding?

4.6 List reasons for adding fillers to both plastics and elastomers.

4.7 When designing a component for mass production in a thermoplastic material, numerous factors must be considered if problems during moulding are to be avoided and if the moulded item is to need only the minimum of machining. In view of these requirements state what you consider to be the most important guidelines when designing for plastics manufacture. Consider both part and die design.

4.8 Why are the bulk of all plastic components made from thermoplastics rather than thermosets, even though the latter produce much stronger parts and are much more tolerant to extremes of environmental temperature?

4.9 Describe, with the aid of simple sketches, the principles of the extrusion processes used for metals and plastics. Only those processes involving production of long lengths need be considered.

4.10 From the production engineer's point of view, describe the major differences between the extrusion of metals and plastics.

4.11 How do you think the profile of an injection moulder's screw feeder is likely to vary from the point where raw material is introduced to where it delivers the hot plastic to the injection nozzle? Indicate the function of each section change that you propose.

4.12 Sketch examples of three types of metal insert commonly used in moulded plastic components, briefly explaining the use of each. How would you ensure that the inserts do not come loose or pull out of the plastic in service?

4.13 Why do dies used in both vacuum forming and blow moulding have water cooling facilities? Why do some components not require such cooling?

4.14 What are composites and why are they increasingly being used in applications that have traditionally been the exclusive domain of metals? Are there any areas where you consider that composites are unlikely to ever replace metals?

4.15 Define the terms elastomer and rubber. Are all rubbers elastomers? What sets elastomers apart from all other engineering materials?

4.16 Why were synthetic elastomers developed and why do their uses now greatly outweigh that of natural rubber even though many applications are still best served by NR?

4.17 Suggest at least three examples of where your material selection would be an elastomer rather than a plastic. Give reasons why you would make such a choice.

Chapter Five
Joining processes

After reading this chapter you should understand:
- (a) how metals and non-metals are joined using processes involving the application of heat
- (b) how metals and non-metals may be joined using non-thermal processes
- (c) the distinction between fusion and solid-state welding
- (d) the differences between the principal types of fusion weld and where they are normally used
- (e) the basic types of joint used in fusion welding and their main areas of application
- (f) why welded structures are prone to distortion
- (g) how the weldability of steels may be assessed
- (h) how plastics are welded
- (i) the fundamental principle of brazing, how it differs from welding and where it is primarily used
- (j) the process of soldering, where it is used and its advantages and limitations as a joining process
- (k) how to join metals and non-metals with adhesives and what constitutes good and bad adhesive joint design
- (l) when and where to use mechanical fastenings to join component parts of an assembly.

Joining things together

Most products are made by assembling a number of component parts, and many of these constituent items need to be joined together either permanently or in such a way that they can be subsequent taken apart again for maintenance or repair purposes. It is therefore essential that today's design engineers are aware of the extensive range of joining methods readily available to modern industry and understand the fundamental principles involved in each process. The principal advantages and limitations of each technology should also be appreciated and joints designed accordingly.

Tradition dies hard, and it is often all too easy to follow the joining methods used successfully in the past without giving due consideration to the more modern, economically attractive, options now available.

5.1 The principal joining processes

The main joining processes used in manufacturing are welding, brazing, soldering, adhesive bonding and mechanical fastenings.

Exercise Before reading further, list as many different joint applications as you can think of (from mending a broken cup to welding together the hull of a ship), stating which joining process you think each employed. Carefully save your list until you have worked through this chapter, as Question 5.1 is based upon its use.

5.2 Welding processes

When one wishes to join together two or more parts permanently with a joint having strength at least equal to that of the material(s) from which the constituent parts are made, some form of welding is normally required. While a large number of welding methods are available, the majority of applications are well catered for by a relatively small group of processes.

There are many general textbooks available (Niebel 1989, DeGarmo 1989, Gourd 1986) that include descriptions of the more specialized welding techniques, but the processes listed in Table 5.1 have been found to satisfy the bulk of manufacturing industry's day-to-day needs. They may be conveniently divided into four family groups, each being distinguished by the method of heat generation used to sustain the localized area of very high temperature necessary in most welding processes. The groups are electric arc, electrical resistance, gas and solid state welding, with the first three being collectively referred to as fusion welding processes.

Table 5.1 Most common forms of welding

Electric arc welding	Electrical resistance welding
Flux-shielded manual metal arc	Spot
Flux-shielded submerged arc (sub-arc)	Seam
Gas-shielded tungsten inert gas (TIG)	Projection
Gas-shielded metal inert gas (MIG)	Flash butt
Plasma arc	
Laser beam	
Electron beam	

Gas welding	Solid state welding
Oxy-fuel gas	Friction
	Ultrasonic

Fusion welding, in its simplest form, involves melting the parts to be joined in the localized area where they meet, and then letting the molten metal mixture within the joint area resolidify to form a metallurgically homogeneous (uniform) joint. In gas welding, and most arc welding processes, an additional element called filler metal is usually required; this is discussed in §5.2.1.

5.2.1 Arc welding – the most common fusion welding process

To melt metals, particularly steels, most of which have melting points in the 1500–1550°C region, requires a large energy input. Fortunately, welding usually only requires that a small area be melted at any instant, and therefore an energy source of some kind can be directed into the localized area to be melted. This allows a progressive welding operation to be sustained without the consumption of excessive power.

One such high-energy source is produced by harnessing the basic characteristics of an electric arc. An arc is generated when an electrical discharge occurs between two adjacent, but not actually touching, metal objects. This discharge is sustained through a path of ionized gaseous particles called a plasma, the temperature generated being over 15000°C inside and in excess of 10000°C at the surface of the arc. Electrical currents of 1000 amps and more are used, although voltages are low and in the range of 30–80 V AC.

The simplest form of arc welding is the deposition of one metal onto the surface of a similar metal. The arc is struck between the surface onto which new metal is to be deposited (i.e. the parent metal) and the source of the new metal. The supply of new metal, or filler metal as it is called, is usually introduced into the weld area in circular rod or wire form and is referred to as a filler rod or consumable electrode (Fig. 5.1).

A major problem with melting and resolidifying metals in air is that molten metals become metallurgically contaminated when exposed to the atmosphere. To overcome this difficulty, either arc welding must be carried out in a vacuum or the molten weld pool area must be enveloped in a protective gas shield to exclude the surrounding air until it has had time to solidify and cool.

In simple arc welding of the form illustrated in Figure 5.1, a gaseous blanket is generated by the presence of a special granular powder covering called flux baked onto the outside of the filler rod. When both filler metal and flux progressively melt in the electric arc, the flux releases both an inert gas and a liquid that floats on top of the molten metal, acting as a secondary air exclusion barrier until solidification of the metal occurs. The solidified flux layer that forms on the surface of the weld is termed slag and must be removed by mechanical means when the weld metal has cooled.

Exercise Can you see the similarity between slag formed in welding and slag that forms on the top of molten metal before casting? If not, reread §2.4.1.4 (*Porosity and sponginess*).

Figure 5.1 The arc welding process. (Courtesy of TWI, Abington, Cambridge.)

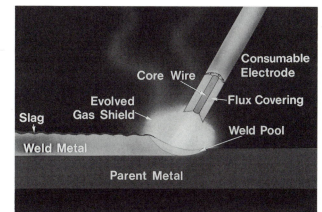

Flux also helps the welding process in other ways, the most important being to enhance the stability of the arc as welding progresses: if constant welding conditions, such as a consistent intensity of arc, are not maintained, weld quality will suffer.

Another important term used in welding is heat-affected zone (HAZ). This refers to the area surrounding the region of the weld whose microstructure has been affected by the heat generated by the welding operation. Ideally, the HAZ should be as small as possible: the hotter a structure becomes during welding the larger the HAZ and the more distortion problems are likely to occur (see §5.3)

Safety warning   As its name implies, arc welding is associated with the generation of a strong electric arc. Such arcs are *very* dangerous: they can cause temporary blindness if viewed with the naked eye for even a short time, and continued exposure will result in permanent eye damage. It is therefore *essential when in the vicinity of arc welding that appropriate eye protection be worn at all times.*

5.2.1.1 Flux-shielded metal arc (sometimes called "manual metal arc" or "stick" welding)

Most welding operations are not used to increase the thickness of a piece of metal, as illustrated in Figure 5.1, but rather to join two or more separate items together. The most common of such arc welding processes is the flux-shielded metal arc process (Fig. 5.2).

A flux-coated consumable electrode is clamped in a hand-held holder and the workpiece is connected to the power supply by a suitable (high-current, low-voltage) cable. The welder strikes an arc between the filler rod and the work surface, and then slowly traverses the rod along the weld path, ensuring that a constant arc gap is maintained at all times, despite a continually reducing rod length – not an easy task! When the rod becomes too short for further use (about 75 mm), the welding operation must stop and a new rod placed in the holder.

This form of welding is the most common arc welding process used because of its flexibility and its ability to be used in difficult locations. However, it does rely heavily upon the skill of the welder.

5.2.1.2 Flux-shielded submerged arc welding

While manual metal arc welding is a flexible general-purpose method of joining mainly ferrous components, there are three fundamental requirements that are not so easy to achieve manually if consistent, high-quality welds are to be guaranteed: (a) the gap between workpiece and filler rod, across which the arc is struck, must be kept constant at all times; (b) welding speed must not vary; (c) there should ideally be only one start and stop throughout an entire weld run, i.e. at the beginning and at the end.

It can readily be appreciated that (a) and (b) are difficult to achieve manually, especially when one considers that the consumable electrode (the welding rod) is become increasingly shorter as the weld progresses and that the operator must constantly view the work through darkened eye protection. As there is a practical limit to the length of welding rod that a welder can comfortably manipulate (about 450 mm depending upon rod diameter), (c) is impossible to achieve if the weld run is of an appreciable length. Clearly there must always be a start and finish to every weld run, but when more than one rod is required to complete the weld, another "start" and "stop" sequence is introduced each time a new rod is used and this increases the risk of weld defects.

To overcome all of the above difficulties a mechanized form of arc welding is available, known as flux-shielded submerged arc (or sub-arc) welding (Fig. 5.3).

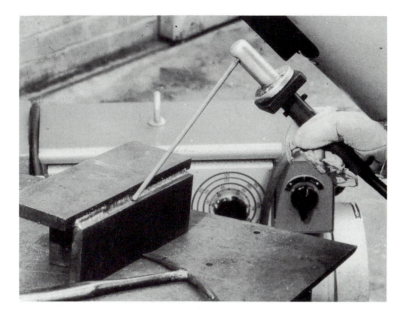

Figure 5.2   Flux-shielded manual metal arc welding. (Courtesy of TWI, Abington, Cambridge.)

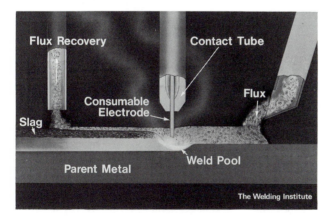

Figure 5.3   Submerged arc welding. (Courtesy of TWI, Abington, Cambridge.)

The welder's hand is replaced by a mechanically propelled welding head through which is fed a continuous supply of electrode wire from a large storage reel, thus avoiding the need to replace rods during a weld run. Head traverse rate is set at a suitable level, and the electrode feed rate is continually controlled to ensure a constant arc gap. This is achieved by a self-regulating controller that continuously monitors the voltage across the arc and compares it with the preset value. Only the minimum voltage necessary to jump the arc gap is generated, so if the voltage value drops the gap must be getting smaller and the electrode wire feed rate must therefore be reduced; conversely, if the voltage rises the gap must be increasing so electrode feed rate must be increased.

This mechanized solution to the problems of manual arc welding, while offering extremely consistent quality of welds, still leaves one practical point to be resolved. By storing the filler rod in one long continuous length on a drum, it is no longer practical to have it covered with the usual flux coating. Not only would the drum have to be enormous to cater for even a modest length of coated electrode, but bending the electrode around the circumference of a storage drum would cause the flux to flake off. To overcome this problem the flux is supplied in granular form from a hopper via a small feeder chute mounted on the front of welding head (Fig. 5.3). Thus, the weld area is provided with a continuous supply of flux, and because the arc operates under this granular flux layer little, if any, arcing is visible during welding – hence the process is called submerged arc.

Sub-arc welding is ideally suited to the production of high-quality welds required in the construction of pressure vessels, and at the same time offers relatively high weld deposition rates.

Often it is more practical to traverse the workpiece under a stationary welding head to achieve the required relative movement between job and welding head. Circumferential welds are best performed this way (Fig. 5.4).

Exercise Try to imagine what is going on within the molten area created by manual metal arc or submerged arc welding. All mixed together are molten metal from each of the parts to be joined, melted filler metal and liquid and gaseous flux. With this in mind try to think of reasons why one of the commonest defects with flux-shielded welding is that of "slag inclusions", i.e. bits of slag trapped within the weld when it solidifies.

Hint: The density of liquid flux is less than that of liquid metal.

5.2.1.3 Tungsten arc (TIG) and gas-shielded metal arc (MIG) welding

A disadvantage of both manual and submerged arc welding is that the flux used leaves each weld run with a surface coating of slag that must be removed mechanically. While flux is required for various reasons, its prime function is to create a gaseous blanket over the weld to prevent atmospheric contamination of the weld metal during its solidification. If this could be achieved by a method that did not leave this unwanted residual slag layer, a significant reduction in the overall cost of weld production would result, and inherently cleaner welds would be produced. Such a solution has been devised, and involves replacing the flux supply with a continuous blanket of inert gas such as a mixture of CO_2 and argon (for low-carbon steels) and helium or argon (for stainless steels, aluminium and other non-ferrous metals).

The two most common arc welding processes that use the inert gas blanket principle are gas-shielded tungsten inert gas welding (TIG) and gas-shielded metal inert gas welding (MIG). Both of these processes are considered simultaneously as their operating principle is identical, the only difference being that MIG welding uses a consumable electrode while TIG welding does not.

In MIG welding the inert gas and filler metal are both fed through the welding head (gun) (Fig. 5.5b). With TIG welding only the inert gas is fed through the gun, the filler metal being manually introduced into the weld area in the same way as in manual metal arc welding (Fig. 5.5a).

TIG welding has an advantage over MIG in that it can successfully weld most ferrous and non-ferrous metals; in addition, as filler metal is not transferred across the arc gap, little if any weld spatter (unwanted small droplets of weld metal) is produced. This, coupled with the absence of any slag, produces extremely clean, high-quality welds, and despite its slowness makes it a popular welding method in the nuclear, chemical plant, aircraft and food industries.

108

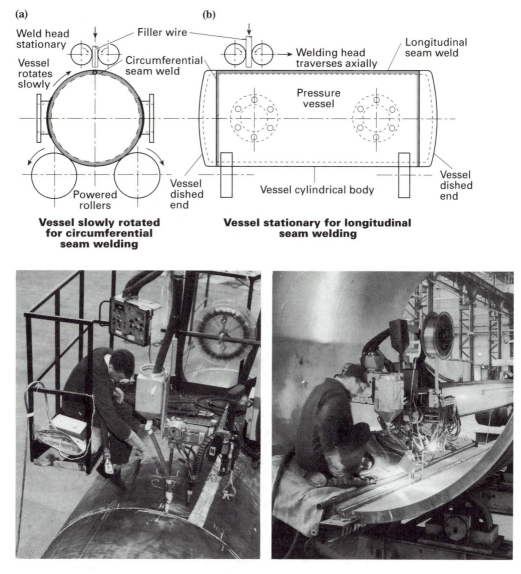

Figure 5.4 Circumferential and longitudinal welding of a pressure vessel. (Courtesy of Babcock Energy Ltd, Renfrew)

MIG welding is a faster process than TIG, and is particularly suited to light to medium steel fabrication work when high production rates are required.

5.2.1.4 Plasma arc welding

This process is a derivative of TIG welding. However, the arc is made to pass through a physical restriction in the form of a small hole before reaching the workpiece. This is achieved by surrounding the non-consumable electrode with a nozzle that has a small central orifice. Inert gas is again used, and is passed both through the central orifice, where it is ionized (the orifice gas), and through the annular outer section of the nozzle head (the shielding gas) (Fig. 5.6).

Passing the arc through such a restriction speeds up the stream of ionized gas in a plasma,

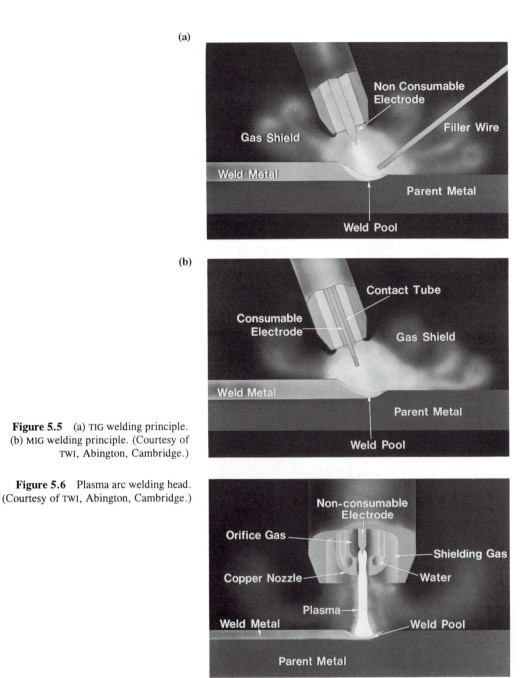

Figure 5.5 (a) TIG welding principle. (b) MIG welding principle. (Courtesy of TWI, Abington, Cambridge.)

Figure 5.6 Plasma arc welding head. (Courtesy of TWI, Abington, Cambridge.)

and this then generates a small area of very high temperature at the work surface. Water cooling of the whole nozzle head is normally required.

Note Plasma is the name given to the glowing, ionized gas that results from heating a material to extremely high temperature. It is composed of free electrons. When in a plasma state, a gas becomes electrically conductive, and can sustain temperatures in the centre of a plasma arc in excess of 30 000 °C (Crafer & Oakley 1992).

110

One of the great advantages of plasma arc welding is that, as only a very small area of precisely directed intense heat is generated, the weld's HAZ is also small. In addition, as this type of heat source is capable of deep penetration, keyhole welding techniques (welds with depth to width ratio of up to 20:1) are possible.

Any metal weldable by the TIG process can also be welded by the plasma arc process, with increases in welding speed of up to 100 per cent being possible.

Low-power plasma arc welding (using currents of 0.1–10 amps) is known as microplasma welding and is particularly suitable for joining foil and thin sheet components.

5.2.1.5 Electron beam welding

The energy source for this type of fusion welding process is generated by accelerating a stream of electrons emitted from a high-voltage source (up to 150 kV) electron gun. The resulting high velocity 0.5–1 mm diameter beam can be precisely focused onto the workpiece, resulting in the melting of its surface. Unfortunately, normal atmosphere slows up the electron stream to such an extent as to make the process ineffective, and it therefore has to be carried out under a high degree of vacuum. However, a benefit of this is that no flux is needed (there is no air present to contaminate the weld metal) and extremely high-purity welds are produced. No filler metal is required because the metal in the joint area is melted in the form of a hole by the circular electron beam. The molten metal from this hole then flows back into the joint area as the weld progresses, creating a deep, narrow weld with little or no distortion (Fig. 5.7).

Most metals can be welded by this process, and even joints involving dissimilar metals can be welded – *generally a practice to be avoided at all costs in most forms of welding.*

Typical thicknesses of steel that can be welded are from 0.1 to 200 mm, and welding speeds up to 5 m/min are possible depending upon metal thickness and maximum beam power available. However, most applications fall within the 0.1–25 mm thickness range (Sanderson 1993).

For practical reasons it is usual for the beam generator to be stationary and the work surface to be traversed under the beam.

Recent developments make it no longer necessary to enclose the whole component in a vacuum chamber. Electron beam units are now available that progressively seal a small section around the weld area in a vacuum while leaving the remainder of the work surface exposed to normal atmospheric pressure. This dramatically reduces the size of vacuum chamber required and the time it takes to re-establish a vacuum after each new component has been loaded in the welder.

Electron beam welding is principally used for high-quality applications typical of the aircraft and aerospace industry, although it is also now used in mass production environments such as

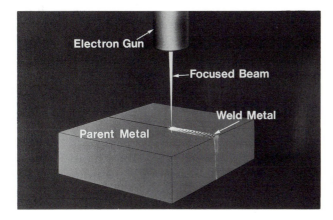

Figure 5.7 Electron beam welding principle. (Courtesy of TWI, Abington, Cambridge.)

the automotive industry, as well as for any application demanding deep-penetration, single-pass, high-quality welds generating a minimal HAZ – typically only 2–3 per cent of that produced by arc welding processes.

5.2.1.6 Laser beam welding

There are a number of similarities between electron beam and laser beam welding, the most obvious being that the thermal energy required for fusion welding is achieved by a high-energy source precisely focused onto the workpiece surface (Fig. 5.8). Welding action is similar to that described in §5.2.1.5, and filler metal is therefore not normally required.

Modern laser beam generators produce a beam of substantially parallel monochromatic (one wavelength) light via either a CO_2 gaseous laser or a neodymium-doped yttrium aluminium garnet (Nd:YAG) solid laser. While the gas laser is the most popular one in laser cutting (see §8.9.4.3), over 70 per cent of laser welding is now Nd:YAG based (Crafer 1992). Furthermore, the YAG laser beam is usually delivered to the weld surface by transmission through a fibreoptic cable system, as this gives very flexible three-dimensional access to the weld area, particularly when mounted on a robotic manipulator arm (Fig. 5.9).

A great advantage of laser beam welding compared with electron beam is that the process does not need to be carried out under vacuum, although a shielding gas such as argon is normally required.

Operating in pulsed mode, typical average power ranges for CO_2 lasers are from 50 W to 5 kW, and for YAG solid state lasers between 50 W and 2 kW. Continuous welding can be as fast as 10 m/min, depending upon laser power and weld parameters, but joint thickness is limited at present to 25 mm, although most applications involve materials less than 1 mm thick.

Laser welding equipment is expensive, a typical CO_2 installation costing over £250 000, so it is necessary to have a high production throughput of high-quality joints to justify such capital outlay.

Safety warning   Both electron beam and laser welding processes generate dangerous levels of electromagnetic radiation. While all commercial machines are well shielded to protect the operator, care must always be exercised by personnel using such equipment to ensure that the safety procedures specified by the equipment manufacturers are adhered to at all times.

5.2.2 The resistance welding principle

In common with the arc welding processes already described, all other fusion welding processes require sufficient heat to be generated in the weld area to create a satisfactory joint. Resistance welding is no different. In this process thermal energy is obtained from the electrical resistance heating principle: all metals have a finite electrical resistance that resists the flow of electric current and in doing so generates heat.

The amount of heat generated (H) depends upon the magnitude of current (I), the electrical resistance of the current-conducting path (R) and the length of time (t) that the current is allowed to flow. The algebraic relationship between these parameters is:

$$H = I^2 R t$$

Even when two pieces of metal are held together under very firm pressure, the electrical resistance at their interface is always much greater than through the metal itself. Therefore,

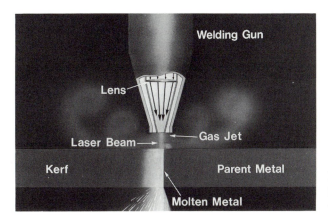

Figure 5.8 Laser beam welding principle. (Courtesy of TWI, Abington, Cambridge.)

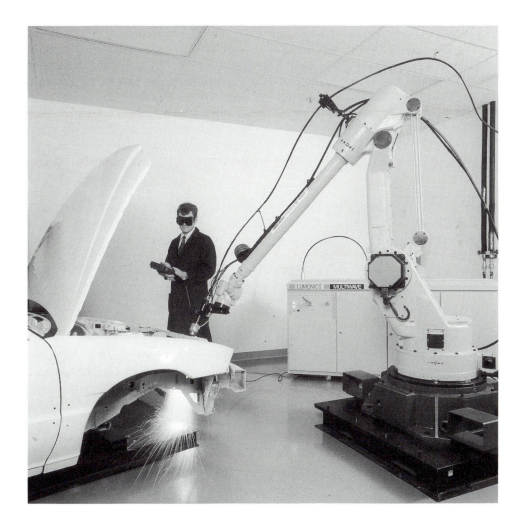

Figure 5.9 Nd:YAG laser robotic welding using a fibreoptic link. (Courtesy of Lumonics Ltd, Rugby.)

most resistance heating occurs at the joint interface, which is fortunately precisely where it is required. Providing adequate heat is developed at the joint, the metals to be welded together will partly melt (*coalesce*), and with the help of the compressive force applied normal to the joint line, a good weld will be produced.

The power generation necessary at the instant of welding is so high (up to 100 kV A) that it can only be sustained for milliseconds. It is therefore not possible to use resistance heating to generate true continuous welds and the technique is restricted to the production of a succession of individual small areas, or spots, of weld. This explains the popularity of resistance welding in sheet metal industries.

No filler metal is required in resistance welding processes, which is just as well, because feeding filler metal into the actual weld area would be impossible in most cases.

5.2.2.1 Spot welding

Spot welding is the most common form of resistance welding and is used principally to join sheet materials (especially low-carbon steel) together. It is therefore not surprising that the biggest users of this process are the car industry and domestic white goods (refrigerators, washing machines, cookers, etc.) manufacturers.

The process of producing a spot weld involves clamping two or more pieces of sheet metal firmly between two water-cooled copper electrodes (Fig. 5.10), and then passing adequate current through this metal sandwich for a sufficient period to generate the required spot weld – the weld nugget (Fig. 5.11). Normally a succession of such spot welds are produced at appropriate

Figure 5.10 Resistance welding machine (vertical arrows show directions of force). (Courtesy of TWI, Abington, Cambridge.)

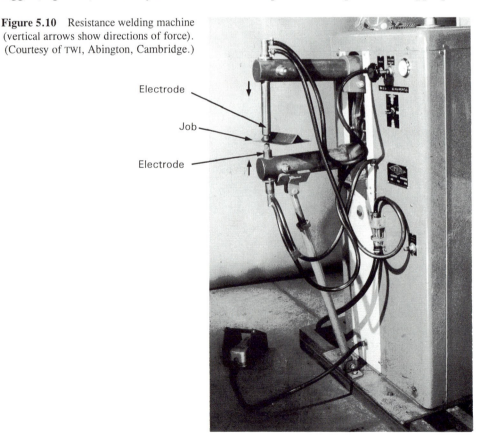

Electrode

Job

Electrode

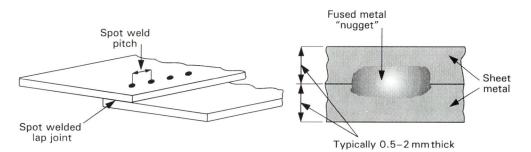

Figure 5.11   Spot welding principle.

intervals (pitch) along the length of the weld line, their size being a function of electrode diameter – typically in the region of 2–10 mm.

Exercise   Why do you think welding copper sheets together is difficult using spot welding?
Hint: What are the electrodes usually made of?

5.2.2.2 Seam welding

Seam welding is similar to spot welding, but instead of using two normal cylindrical copper electrodes, the workpieces are passed between copper alloy wheels or rollers that serve as the two electrodes. The power supply is then pulsed at an appropriate frequency which, when coupled to a suitable workpiece linear traverse rate (controlled by the electrode wheels' rotational speed), creates a series of overlapping spot welds that resembles stitching (Fig. 5.12).

Providing the distance between one spot weld and the next is small enough, each individual weld nugget will tend to slightly overlap, so producing what is, in effect, a continuous weld. Because of this weld continuity, seam welding is successfully used to produce, at high speed, liquid- and gas-tight joints in the can and drum industry.

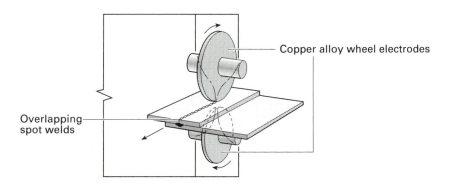

Figure 5.12   Seam welding principle.

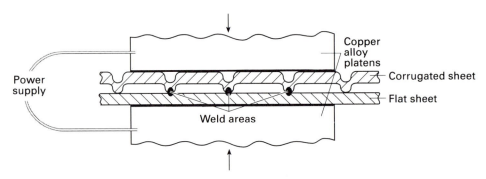

Figure 5.13 One form of projection welding.

5.2.2.3 Projection welding

This is another variant of spot welding but, unlike spot and seam welding, copper electrodes are no longer used to concentrate the electrical current path.

There are various forms of projection welding, and a typical one is illustrated in Figure 5.13. Here small projections are raised on one of the sheets or plates to be joined, so presenting a series of parallel current flow paths to the electrical supply. During the welding process, these small projections are pressed against the mating flat sheet by a pair of copper platens (flat plates), which are also used to transmit the electrical supply. This results in all the projection areas held between the platens being welded simultaneously. However, maximum power availability must be considered if the number of weld areas is large.

Other applications include the production of wire mesh concrete-reinforcing mats, welding occurring at the intersection of each of the reinforcing wires, and the welding of captive nuts onto sheet metal components.

5.2.2.4 Flash butt welding

While still relying upon the resistance heating principle (§5.2.2), flash butt welding also involves a preliminary arc welding stage (Fig. 5.14). The electrodes used are in the form of a pair of electrically conductive clamps, which are used to grip firmly the two pieces of metal to be joined, and to conduct the high electrical current required (up to 100 000 amps) to the weld interface.

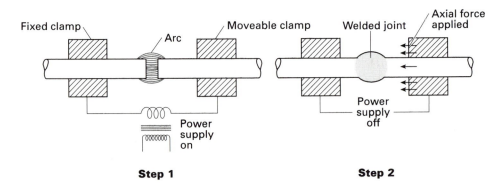

Figure 5.14 Flash butt welding principle.

The two halves of the joint are brought into close proximity to each other, but not touching, and the electrical power supply is switched on. Arcing between the two joint halves occurs (Fig. 5.14, step 1), and continues until the whole joint area is at the required temperature. Power is then turned off and the movable clamp is moved rapidly towards the fixed one, thus pushing the two halves of the joint together under heavy pressure. Molten metal is forced out of the joint area (Fig. 5.14, step 2) and, when the joint has cooled, this must be removed by machining.

This process is ideal for joining bar, tube and other section formats, and is therefore frequently used for joining boiler tubes and rail track, and in metal window frame assembly

5.2.3 Gas welding

Gas welding usually refers to the use of an oxy-acetylene flame as the thermal source required for welding, and involves the burning of acetylene (the fuel gas) in an oxygen-enriched atmosphere to create a high-temperature flame. This flame is generated by mixing the two gases in a hand-held welding torch fitted with interchangeable nozzles to enable the flame size to be altered to suit the requirements of each welding job.

(a)

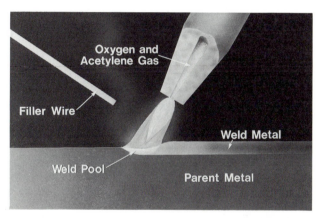

(b)

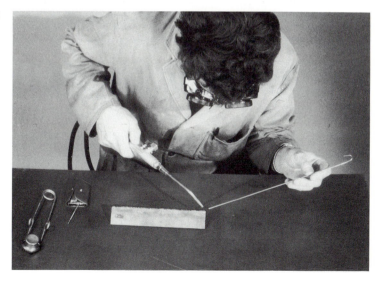

Figure 5.15 (a) Oxy-gas welding principle. (b) Oxy-gas welding in action. (Courtesy of TWI, Abington, Cambridge.)

As with other fusion welding methods, the thermal source (the flame) is used to heat and melt both parent and filler metals in the joint area (Fig. 5.15).

Many jobs that, in the past, would have been gas welded are now carried out using much faster arc and resistance welding processes. Nevertheless, it remains an inexpensive, very flexible and portable process (no external electrical supply is required), and is particularly suitable for many maintenance welding jobs, for the manufacture of sheet metal ducting and pipework in the heating and ventilating industry and for motor vehicle repair work. Furthermore, it is very convenient to have immediately to hand a gas flame for other day-to-day operations such as flame cutting, brazing (§5.6) and soldering (§5.7).

5.2.4 Solid state welding

There are a small number of welding processes that do not depend upon gas flame, electric arc or resistance heating to generate the temperatures necessary to produce a fusion welded joint. Nevertheless, the joint area still has to be heated if welding is to be achieved. Two such processes of increasing industrial importance are friction welding and ultrasonic welding.

5.2.4.1 Friction welding

In friction welding the thermal energy needed to create a weld successfully is produced by mechanical rubbing friction between the two parts to be joined. This is achieved by rotating one half of the joint while the other half is held stationary, except for being pushed axially against the rotating part (Fig. 5.16a2). When the joint area has been heated to a plasticized state, rotation is stopped, and to complete the weld the two parts are even more firmly axially pressed together – typically with a force of 10 MN (Fig. 5.16a4).

Note   This is a similar welding process to flash butt welding (5.2.2.4) except that frictional heat is used instead of an electric arc.

Although this process is limited to joints of which at least one of the two parts must be rotatable in either a chuck or other gripping fixture, it does have the great advantage that dissimilar metals can be readily joined. For small 1–2 mm diameter components, rotational speeds up to 80 000 rev/min are necessary, but for much larger parts speeds may be as low as 50 rev/min.

Weld quality is both good and consistent across the whole joint area.

By automating component handling and the welding cycle (which can even incorporate automatic machine removal of excess metal around the outside of the joint), productivity is now sufficiently high to make friction welding a widely used process, particularly in the automotive industry, where it is used in the production of such items as axle casings, transmission shafts and turbocharger internals.

Even thermoplastics can be joined using friction welding and, as energy requirements are small compared with arc and resistance welding, the HAZ is also small.

5.2.4.2 Ultrasonic welding

While also relying upon a form of frictional heat, ultrasonic welding is somewhat unusual in that it uses an electrical transducer to establish a resonant vibration at the interface of the two parts to be joined. The parts are trapped between an oscillating tip and the welding machine's

(a)

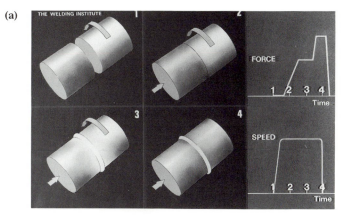

(b)

Figure 5.16 (a) Friction welding principle. (b) Completed joint before removal of excess material. (Courtesy of TWI, Abington, Cambridge.)

Excess metal

work table (the anvil), and the resonance is sufficiently intense (up to 200 kHz) to generate enough frictional heat to create the required weld – a form of spot weld (Fig. 5.17).

Welds are completed in a second or less, although this is largely due to the low thermal capacities of the thin foil, sheet, wire and plastic materials that this process is used to weld.

Like friction welding, this process can be used for joining both metals and plastics.

5.2.5 Types of fusion weld

The three most common forms of fusion weld are bead, butt and fillet (Fig. 5.18).

5.2.5.1 Bead welds

Bead welds are only suitable for joining thin sheet metal and for deposition of hard facing materials such as Stellite. Machining of the joint's edges preparatory to welding (usually called

119

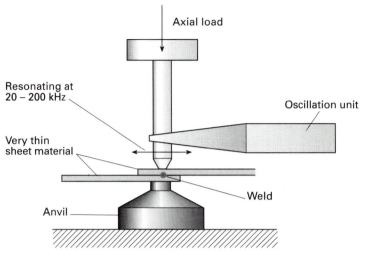

Figure 5.17 Ultrasonic welding.

weld prepping) is not appropriate with this form of weld because of the thinness of the sheet metal normally involved.

5.2.5.2 Butt welds

These are used when thicker metals are to be joined and are typical of the main seam welds required in pressure vessel manufacture (Fig. 5.4).

These joints are invariably far too thick to be produced using just one weld run, and usually require a number of successive weld layers to build up the full weld thickness required (multirun welds). Figure 5.18b illustrates this, but in reality, of course, all weld runs and the joint areas of the parts to be joined fuse together to form a completely homogeneous fusion joint.

Exercise   Since multirun welds are frequently produced by the sub-arc process (§5.2.1.2), consider what work is involved *between each weld run*, and how this affects the overall cost of producing this form of the welded joint.   *Hint*: Sub-arc welding is a flux-shielded process!

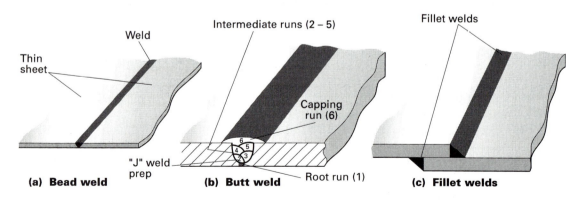

Figure 5.18 Types of fusion weld.

Joint preparation involving accurate edge profile machining to ensure uniformity of the weld across the entire thickness and length of the joint is normally essential (Fig. 5.18b). The shape of weld prep used is a function of material thickness and the welding process employed. Welding design manuals and the appropriate British Standard (British Standards Institution 1984) specify which weld prep design should be used for any given application.

5.2.5.3 Fillet welds

Fillet welds are used for making tee and lap joint configurations (Fig. 5.19), typical of structural steel work, and are also used in corner joints. As with butt welds, larger fillet welds usually require multiruns. No edge weld preparation is necessary because, by the nature of the angular joint shape, open access is available across the whole weld thickness.

Unlike bead and butt welding, fillet welds do not necessarily need to be continuous over the whole length of the joint. On occasion, intermittent lengths of fillet weld are adequate, thus saving both time and weld materials.

5.2.6 Factors that commonly affect fusion weld quality

Consistency of welding conditions is an essential factor in controlling the quality of any weld. This refers not only to the process parameters and equipment controlled by the operator, but also to the environment in which welding is being carried out. For example, if welds are being laid down adjacent to a frequently opened external door, this can thwart the efforts of the best welder to produce uniformly acceptable results.

In the case of manual metal arc welding it is vital that the consumable electrodes used are not damp, as this will cause metallurgical and gaseous inclusion problems in the welds produced. Most welding shops preheat rods before use to avoid such quality problems.

For all but the most general-purpose welds, only welders whose competence to carry out specific types of welding is checked regularly should be allowed to carry out high-quality welding work. This is usually mandatory in pressure vessel manufacture.

With flux-shielded welding methods, such as manual metal arc and submerged arc, much of the total time required to produce multirun welds is taken up in the meticulous removal of all traces of slag left from the previous weld run before the next run can be laid down. Although time-consuming, this step is vital as slag inclusions in the finished weld necessitate gouging out the weld area affected and replacing it. This is no easy task and can usually only be carried out twice in any one area before unacceptable metallurgical changes occur in the surrounding HAZ.

Welds should be as strong as the parent metal being joined, but this will only be so if the filler metal (when used) is of at least the same metallurgical specification as the metal being welded. Also, joints will not attain their full strength if complete fusion of the filler metal and the parts being joined does not occur. Such lack of fusion can arise for numerous reasons, but the most common cause is insufficient heating of the weld area. This can also cause incomplete penetration across the full thickness of the weld.

Not only must a weld be right, but it must also look right. It is therefore important that a finished weld has the correct shape, is as uniform as possible and has all traces of slag removed. With multirun welds, surface finish is only of particular importance in the final, or *capping*, run (Fig. 5.18b), as the better the finish of this run, the less time has to be spent in final dressing with a hand grinder (Fig. 6.7).

Other defects such as porosity, cracking and weld corrosion also occur from time to time, but possible causes are too complex to be considered here and are discussed in detail in various texts published by specialist welding organizations such as The Welding Institute (TWI).

5.2.7 Types of fusion joint

Five basic types of joint cover most fusion welding applications, and these are termed butt, tee, corner, lap and edge joints.

5.2.7.1 Butt joints
Butt joints are used to weld plate in the same plane and employ either bead or multirun butt welds, depending upon joint thickness (Fig. 5.18).

5.2.7.2 Tee joints
These are used to join two items to one another at right angles. Fillet welds are used on both sides of the tee junction. Tee joints are very susceptible to welding distortion, but this can sometimes be minimized by welding both sides of the joint simultaneously.

5.2.7.3 Corner joints
These are similar to one half of a tee joint, fillet welds again being used.

5.2.7.4 Lap joints
As all lap joints involve the filling in of corner(s) along the joint (Fig. 5.19), fillet welds are used, although resistance seam or spot welding (§5.2.2) is more appropriate in the case of thin sheet metal lap joints.

5.2.7.5 Edge joints
These are mainly joints associated with relatively thin materials, so bead welds are normally used if a continuous weld is necessary. If this is not essential resistance seam or spot welding may be a better option.

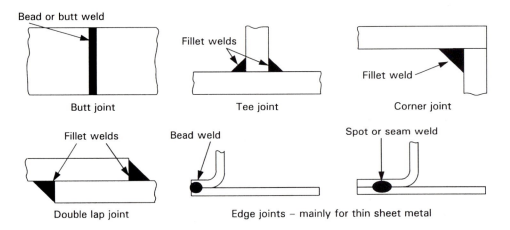

Figure 5.19 Principal forms of fusion welded joint.

5.3 Distortion of welded structures

Probably the greatest problem with welded structures is that of distortion. While it cannot always be avoided, judicious design and manufacturing procedures can certainly minimize its effects.

Distortion is primarily caused by the heating and cooling of welds and their surrounding HAZs as the welding process progresses. The pool of molten weld metal is clearly the hottest area, and it therefore contracts most on cooling. In so doing tensile and torsional stresses are induced in the structure which, if large enough, can cause significant structural distortion.

The use of properly designed welding fixtures to resist such distortion can work, but they also run the risk of locking stresses in the structure, which are then frequently released during subsequent machining or in service. Pressure vessel manufacturers often preheat the main joint areas before welding, and then thermally stress relieve (anneal) the vessel after welding is complete but before machining to minimize the risk of this happening.

While welded structure designers must ensure that the end product is strong enough to do the job, and that the types of weld specified are suitable, they must also pay due consideration to the method of manufacture, and ensure that access to all weld areas is not too difficult. Such forethought also ensures that the structure is designed to minimize manufacturing costs, for example by not welding thick and thin sections together, minimizing the amount of weld prep machining necessary and letting distortion freely occur in areas where subsequent machining is required anyway.

It is worth noting that butt welds cause shrinkage principally normal to the direction of the weld run, and that the more runs used in a multirun joint the greater will be the total distortion that occurs. Therefore only the minimum number of runs should be used, which also makes economic sense as every run costs money, as do unnecessary slag removal operations.

If joint strength is not a major issue, intermittent welds should be considered, as these tend to cause less distortion than continuous welds, providing they are laid down in the correct sequence.

Exercise Did you know that bicycle frames are assembled using resistance and arc welding processes as well as brazing (§5.6)? Take a close look at a typical bicycle and see if you can deduce which joining processes are used where, and why.

5.4 Weldability of steels and cast iron

There are an extremely large number of different steels available, ranging from low-carbon steels to very high-tensile and anticorrosive alloy steels. Unfortunately, not all steels can be successfully welded, and therefore designers must take this factor into consideration when calling for a given steel to be used in a welded structure.

A steel's weldability can be checked in relevant welding handbooks and national standards (British Standards Insititution 1984) or by reference to either filler metal suppliers or to appropriate research/trade associations such as the Welding Institute, Abington, Cambridge.

The adverse effects upon the arc welding process of the various alloying elements commonly added to steel, may be assessed by calculating the carbon equivalent (CE) of the steel in question.

$$\text{CE} = \%\text{C} + \frac{\%\text{El}_1}{n_1} + \frac{\%\text{El}_2}{n_2} + \frac{\%\text{El}_3}{n_3} + \ldots + \frac{\%\text{El}_n}{n_n}$$

where C is carbon, $\text{El}_1, \ldots, \text{El}_n$ are the alloying elements contained in the steel and $n_1, \ldots, n_n$ are the empirical indexes specific to each alloying element (Lindberg 1983, Bailey et al. 1993).

The validity of this empirical (experimentally derived) equation is based upon the fact that, while carbon in steel is the biggest single contributor to weld hydrogen cracking problems, other common alloying elements such as manganese, nickel and chrome also contribute to a greater or lesser degree to the same problem. Surprisingly, molybdenum and vanadium have the reverse effect. The above CE equation can therefore be used to compute the carbon content of a hypothetical plain carbon steel that would have the same welding characteristics as the alloy steel under consideration.

It is generally considered that steels with a CE above 0.4 per cent are difficult to weld without resorting to preheating and, possibly, controlled cooling of the joint area. Above 0.5 per cent special low-hydrogen filler metal rods are essential, and above 0.8 per cent welding becomes extremely difficult, if not impossible, to achieve.

Cast iron is very high in carbon (2–4 per cent). Therefore, according to the above CE equation, and ignoring the effects of any alloying elements used, it should be totally impossible to weld. This is not entirely true however in that, while cast iron fabrications are not made, repairing cast iron castings by fusion welding is possible. Electric arc can be employed, although oxygas welding is preferred, despite the need to preheat the weld area and to ensure that post-weld cooling is as slow as possible. In fact, the weld pool area does not really become liquid as it does when welding steels, but is only heated to a plasticized consistency. Brazing (§5.6), using bronze or nickel-copper filler, is preferred to welding in many cases as high structural strength of the joint is not usually a factor in applications using cast iron castings.

5.5 Welding of plastics

In Chapter 4 it was stated that there are two basic family groups of plastics – thermosetting plastics and thermoplastics. While components made from either of these types of plastic may be bonded together using either a suitable solvent or adhesive (§5.8), or joined using mechanical fasteners (§5.9), welding them is not quite so straightforward.

Because the two basic types of plastic are fundamentally different in nature, only thermoplastics lend themselves to welding techniques as they are the only ones that soften when heated. This is why thermosets must be joined using either mechanical or adhesive joining methods.

Note   If you do not know why this is, please reread Chapter 4, §4.1.1 and §4.1.2.

For small joints a simple soldering iron may be sufficient to preheat the joint area but, whatever heat source is used, the welding of thermoplastics usually involves preheating the surfaces to be joined, removing the heat source and then holding the joint together until it cools. A hot gas torch is commonly used to plasticize (soften) the joint area and filler material may or may not be necessary.

Another method of introducing heat to the joint area is by the use of an electrically heated platen. This is placed between the two flat faces to be joined until they reach the required temperature. The platen is then removed, and the two halves of the joint are pressed together until they cool and the joint is complete (Fig. 5.20).

As is the case when welding metals, most welding techniques used to join thermoplastics only differ in the form and method of application of the thermal energy necessary to produce a satisfactory weld in the joint area. Indeed, a number of the metal welding methods already described are also used to join plastics. Ultrasonic (§5.2.4.2) and friction welding (§5.2.4.1) are two such processes, but in the case of friction welding the constraint that one of the component parts of the joint must be rotatable limits its areas of application.

A radiofrequency (27 MHz) energy source is often employed for welding joints of thin thermoplastic materials (0.15 to 0.5 mm thick), of the type used to prepackage DIY products (Fig. 5.21).

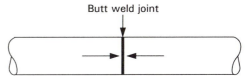

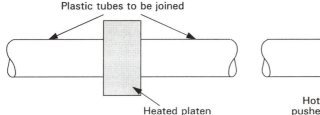

Figure 5.20 Flat plate welding of two plastic pipes.

Figure 5.21 Radiofrequency welding for blister pack sealing. (Courtesy of Stanelco Products Ltd, Fareham.)

5.6 Brazing

Although brazing can be used to join ferrous metals, it is not a substitute for welding. This is because the filler metal used has a much lower melting point and structural strength than ferrous metals. It is therefore mainly used to join non-ferrous metals. Unlike welding, brazing is a metal joining process in which *the parts to be joined do not melt*, and the non-ferrous filler metal used normally melts at a temperature considerably below the melting point of either of the joint's constituent parts.

Molten filler metal, with the assistance of a powder type of flux to help wet the joint surfaces, flows by capillary action between the heated but unmelted joint parts. Filler metal diffuses into the pure base metal(s) of the joint surfaces, and creates thin surface layers that are metallurgically changed because they then comprise an alloy of filler and base metal. It is these alloyed surfaces that act as the bridge between the component parts to be joined and the filler metal (Fig. 5.22).

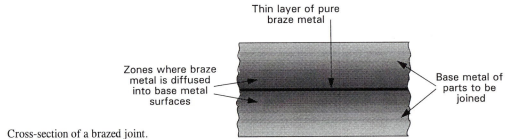

Figure 5.22   Cross-section of a brazed joint.

Brazing is usually carried out in a normal atmosphere, but in the case of aluminium brazing the process is generally performed under vacuum conditions to prevent oxidation problems.

Various heat sources are used, the most common ones being oxy-acetylene flame torch, furnace, electrical resistance and induction heating. Flame heating is the most flexible method, with induction heating offering the best heat control.

Most brazing filler metals fall into one of three types: copper-based alloys, silver-based alloys and aluminium-based alloys. Copper alloys have a melting point range of between 700 and 1150°C and are particularly suitable for brazing steel, cast iron and copper components. Silver-based alloys have a melting point range of between 640 and 850°C and are suitable for brazing materials based on titanium, copper, brass and nickel. (Brazing using silver-based alloys is often also referred to as silver soldering.) Aluminium-based alloys have a melting point range of between 570 and 620°C and are used exclusively for brazing aluminium and its alloys. However, aluminium is an extremely difficult metal to braze as the aluminium filler metal melts at only just below the melting point of the aluminium components to be joined. Induction heating is preferred as this provides extremely precise and accurate temperature control of the heating process.

Note that filler metals that melt below 450°C are classed as solders (see §5.7).

Lap-type joints (Fig. 5.23e) are preferred as they provide excellent strength, whereas butt joints (Fig. 5.23b) are best avoided if possible as the contact area of the joint is small and it is easy to introduce too thick a layer of filler metal – a frequent cause of weak brazed joints (Fig. 5.23a). If a degree of mechanical constraint can be incorporated into the joint design, this further enhances its strength (Fig. 5.23f).

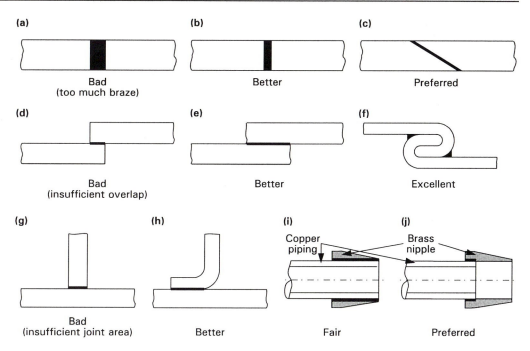

Figure 5.23 Good and not so good brazed joint configurations.

One of the great benefits of brazing is that, with the relatively modest temperatures involved, the HAZ is small, and there is therefore less risk of component distortion or other adverse thermal effects. This is one of the reasons why brazing is popular for the construction of bicycle frames and for car body repair work.

5.7 Soldering

Soldering, or soft soldering as it is often called, is a similar process to brazing except that the filler metal used, the solder, has a much lower melting point and very low tensile strength. Solders are alloys of tin and lead that melt in the range of 175–400°C. Soldered joints must therefore not be placed under stress. Lap joints should be used in preference to butt joints, and service temperatures should not exceed 150°C.

A soldered joint is made by first heating the region of the parts to be joined to a temperature somewhat in excess of the solder's melting point, either with a gentle gas flame, or in an induction coil (Fig. 5.24a) or, more commonly, by an electrically heated soldering iron of suitable power – typically 25–500 W (Fig. 5.24b). Solder is next introduced into the heated zone, where it melts and, under capillary action, then runs into the small gap (0.1–0.15 mm) between the joint members. Upon removal of the heat source the solder resolidifies, so forming the required joint.

As with brazing, flux is used to avoid oxidation of the filler metal and to aid its capillary flow into the joint area. It must be applied to both the previously cleaned joint members and to the solder, although nowadays flux is normally incorporated within the solder.

127

(a)

Figure 5.24 (a) Induction heating coil.
(b) Soldering using an electrical
soldering iron. (Courtesy of Solbraze
Products Ltd, Erith.)

(b)

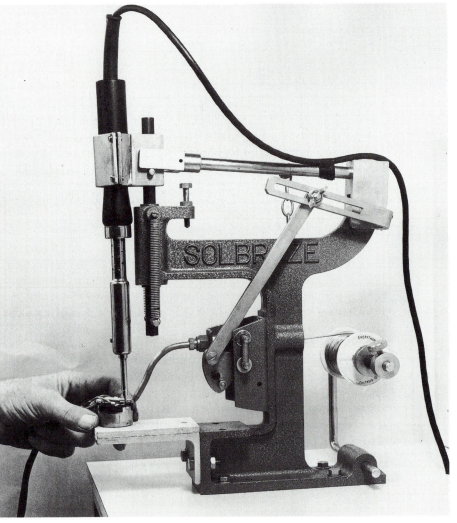

Soldered joints are best suited to copper, tin and brass components, although aluminium- and zinc-based metals can also be soldered, but only under special conditions and with special aluminium/zinc-based solders.

Ferrous metals cannot be soldered.

Unlike welding, and to a lesser degree brazing, a soldered joint possesses little significant structural strength. It is principally a low-temperature joining method that is ideally suited to the needs of the electronics industry because of its ease of use, low cost and reversibility. To solder each of the large number of joints found on modern printed circuit boards individually would take far too long. They are therefore usually mass soldered (Leonida 1981) by dip or flow soldering. In essence, the joints to be soldered are simultaneously dipped into a bath of molten solder (Fig. 5.25). Upon removal excess solder falls back into the bath, leaving the solder in each of the joints to solidify.

Soldering can also be used when liquid- or gas-tight joints are required, providing the metal to be joined is readily solderable and its operating temperature is well below the melting point of the solder used. An example of this type of application is the production of car radiators.

Figure 5.25   Dip soldering machine used in printed circuit board manufacture. (Courtesy of Solbraze Products Ltd, Erith.)

5.8 Adhesive bonding

The use of adhesives to produce well engineered and reliable metallic and non-metallic joints has now been universally accepted, thanks largely to major development work to extend their applications in the aerospace industry. Increasingly, adhesive bonding is being chosen in preference to the more traditional options of soldering, brazing and mechanical fastenings. While organically based adhesives (those derived from plants or animals) are acceptable for joints that are largely non-load bearing (e.g. domestic types of glue), in the engineering field complex chemical adhesives are used as they are capable of significant strength if properly applied.

Irrespective of the adhesive employed or how it is applied, the surfaces to be joined must be scrupulously clean before adhesive application and have a pronounced chemical affinity for the adhesive used. Invariably, a better bond is achieved if the surfaces of the faces to be joined are roughened to provide an improved adhesive key. Too thick a layer is not advisable, and with most adhesives, a 3–4 mm coating is usually sufficient.

A wide range of industrial adhesives are available, but the most common ones are either elastomer or synthetic resin based.

Elastomeric adhesives are normally made from either a rubber or silicone base, and are used for joining non-metallic materials to metal surfaces. However, at least one of the surfaces in the joint must be porous to permit evaporation of the solvent- or water-based carrier used in this type of adhesive. Elastomer adhesives are ideal for many uses, e.g. in the aircraft industry, where rubber's good vibration-damping properties are exploited. However, because of their low tensile strength and poor peeling resistance (Fig. 5.26), their range of applications is very small compared with synthetic resin adhesives.

Figure 5.26 Stress distribution in riveted and bonded joints. (Courtesy of Ciba Polymers, Duxford, Cambridge.)

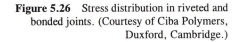
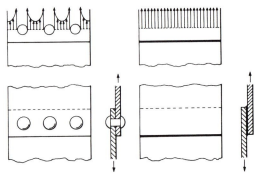

Synthetic resin-based adhesives are most commonly made from either thermoplastic or thermosetting plastics. They are suitable for a wide range of modern structural joint applications, and are used in the bulk of adhesive bonding applications. Thermosetting resins are preferred for high-strength and creep-resistant joints, but because thermoplastic adhesives soften with in-

Figure 5.27 Types of stress set up in adhesive joints. (Courtesy of Ciba Polymers, Duxford, Cambridge.)

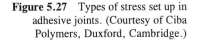

Tension   Compression   Shear

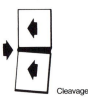

Cleavage   Peel

130

creasing temperature they are usually restricted to relatively low-strength joints applications.

A major drawback of all synthetic joining materials is that their maximum operating temperature is very low. Even the best synthetic-based adhesives are limited to no more than approximately 250–300°C, with few performing well above 150–180°C.

One of the greatest advantages of adhesive bonding is that the loads imposed upon the joint are spread evenly over the whole joint area (Fig. 5.26), but to make optimum use of this characteristic the designer must be familiar with the types of loading and the resultant stresses imposed upon different configurations of bonded joint.

Adhesive joints are loaded primarily in tension and shear, and these types of loadings produce tensile, compressive, shear, cleavage and peel stresses within the joint (Fig. 5.27).

To minimize the risk of creep and peel problems, two fundamental rules in adhesive joint design are to maximize joint area and ensure that the joint line is as parallel to the direction of the applied load as possible. This is why butt joints should be avoided whenever possible (Fig. 5.28).

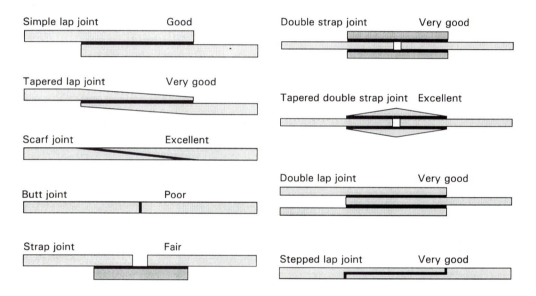

Figure 5.28 Good and not so good adhesive joint configurations. (Courtesy of Ciba Polymers, Duxford, Cambridge.)

As with brazed joints, if the mechanical design of a joint can be configured to assist its inherent strength, so much the better (Fig. 5.29). However, it is important to ensure that the manufacturing costs of such joints do not outweigh the benefits of the increased bond strength achieved by, for example, introducing the need for joint preparatory machining that would otherwise not have been required.

Typical applications of adhesive-bonded joints are brake and clutch lining attachment (rivets are no longer used); lightweight honeycomb panels for internal use within aircraft structures; helicopter rotor blade construction using carbon fibre composite; liquid shakeproof washers (anaerobic adhesives) for screws and nuts; and securing low-stress components in car assembly, especially internal trim.

While almost any combination of materials can be successfully joined, no one universal adhesive exists. The key to guaranteeing the integrity of any adhesive joint is therefore to ensure that the adhesive developed for the particular material(s) being joined is used. There is

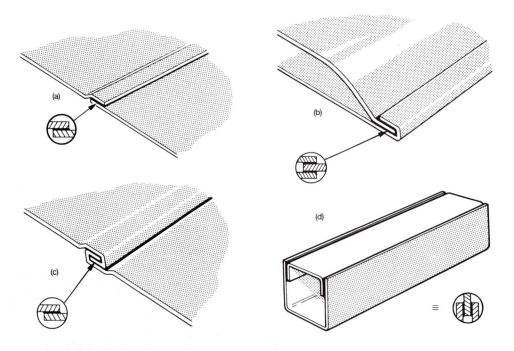

Figure 5.29 Joint designs that enhance their structural strength. (Courtesy of Ciba Polymers, Duxford, Cambridge.)

no better way of guaranteeing this than by contacting major adhesive manufacturers and taking advantage of the wealth of customer service information and advice available from them (e.g. Ciba-Geigy undated).

Exercise   List as many different day-to-day and engineering joint applications that use adhesive bonding as you can think of. Pay particular attention to the interior of any car to which you have access.

5.9 Mechanical fasteners

Fasteners is a generic term covering a wide variety of familiar devices used to join items together, and includes nuts and bolts, screws, rivets, clips, locking pins, wire stitching, stapling, etc. (Fig. 5.30).

Although not strictly fastenings, press and shrink fitting of precision components can also be included within this general grouping.

Mechanical fasteners can be designed to be either permanent fixings or ones that lend themselves to easy removal when required. The design engineer usually knows from experience which form of fastener is most appropriate for a given situation, but with an ever-increasing range of fastenings available it is becoming increasingly difficult to know which one to select. A helpful paper dealing with this problem is Technical Editor (1992).

Permanent fixings are typified by rivets, stapling and metal stitching, while other types of permanent/semipermanent mechanical joining are achieved by interlocking joint designs of

132

Figure 5.30   Range of mechanical fastenings. (Courtesy of Arnold Wragg (Nuts and Bolts) Ltd, Sheffield.)

mating parts which, once assembled, can be made permanent by the application of press pressure to close up the joint (Fig. 5.31a). Alternatively, the use of locking tabs can provide the opportunity for limited joint disassembly, but usually for only a very few times before the tabs break off (Fig. 5.31b).

Non-permanent fastenings include nuts and bolts, screws, retaining clips and rings, locking pins, locknuts and any other device that can be removed and replaced at will. This is by far the most common engineering assembly method used, mainly because there is no restriction on section thicknesses that can be assembled and because of the ease with which joints may be dismantled and then reassembled using only everyday hand tools. A classic example of this

(a)   **(b)**

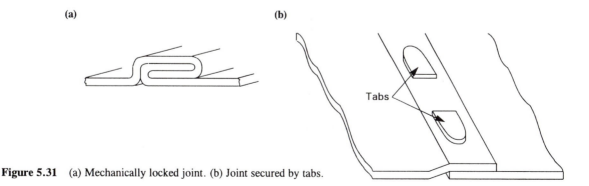

Figure 5.31   (a) Mechanically locked joint. (b) Joint secured by tabs.

approach is the internal combustion engine where the ability to strip down the engine for maintenance purposes is essential. Mechanical fastenings are also used extensively, both externally and internally, in domestic equipment such as video recorders.

When designing joints involving fasteners it is particularly important that environmental service conditions are taken into account and that the parts to be joined and the fasteners used are sufficiently strong to cope with the service loads imposed. The design calculations involved must therefore take full account of the significant localized stress-raising effects caused by most fastenings and by the holes into which they are fitted.

While there is generally no restriction on joining dissimilar materials, care must be taken to ensure that mating materials will not chemically react with one another over time.

Exercise List as many different joint applications that use some form of mechanical fastening as you can think of, differentiating between permanent and non-permanent methods. See if you can work out why the designer chose the method used in each application.

Hint: Do not forget to consider the all-important matters of production volume and costs.

5.10 Summary of the principal characteristics, advantages and disadvantages of joining processes

Process	Characteristics	Advantages	Disadvantages
Welding			
Fusion: arc and oxy-gas heating methods	Coalescence (fusing together) of similar parent metals by heating the joint surfaces, usually with the addition of filler metal	High-strength joint – as strong as, or stronger than, the parent metal. Wide range of flexible processes available, e.g. stick, sub-arc, TIG, MIG, plasma arc, electron and laser beam, oxy-gas	Joint cannot be dismantled. High risk of thermal distortion and stress build-up – may result in the need for expensive stress relief, although not usually necessary with gas welding. Possible adverse metallurgical changes in the HAZ. May need expensive weld prep machining and/ or welding fixtures. Bad practice to weld dissimilar metals or similar metals of widely varying thickness. In most cases weld quality depends greatly upon skill of welder (but not with sub-arc)
Resistance heating methods: spot, seam, projection, flash butt	Fusion process usually not involving actual melting of the surfaces to be joined	Quick process – can be readily automated. High degree of weld	High initial capital cost. Some surface preparation is required but not as

Process	Characteristics	Advantages	Disadvantages
	No filler metal required (or possible) Welds are a series of individual spot welds	consistency Little distortion of the weld area owing to small HAZ	much as with arc or gas welding – usually involves mainly surface cleaning All but flash butt welding really only suitable for lap joints
Solid state: friction, ultrasonic	Heat for welding process generated by creating friction between parts to be joined	Quick, energy-efficient processes, offering consistent quality No filler metal and minimal surface preparation required	With friction welding one of the components must be rotatable and be able to withstand the high torque imposed during the welding process Part sizes limited
Brazing Filler metals based upon: Copper-based alloys; melting point 700–1150°C Silver-based alloys; melting point 640–850°C Aluminium-based alloys; melting point 570–620°C	Joining of components by a filler metal having a melting point significantly lower than the metal(s) being joined, i.e. the joint members are not melted during the joining process The filler metal acts as a metallurgical "bridge" between the constituent parts of the joint	Much less thermal distortion than with welding processes owing to smaller HAZ Dissimilar metals with differing melting points and thicknesses can be joined Complex assemblies can be joined in several steps by using filler metals with progressively lower melting temperatures	Joint strength less than that of parent metal(s) of the joint members Fluxes used to remove oxide films can be corrosive Limited environmental operating temperatures, and not suitable for dynamic loadings above approximately $100\,MN/m^2$ Shear strength of the joint is limited by the shear strength of the filler metal Cost of joint preparation can be significant Aluminium difficult to braze – usually performed under vacuum
Soldering Filler metal is usually a tin/lead alloy; melting point 175–400°C	Similar process to brazing but with an even lower melting point alloy used as the "bridge" between the constituent parts of the joint	Produces gas- and liquid-tight joints quickly and at low cost Joints can be easily disassembled by remelting the solder Ideal for unstressed electrical connections No expensive equipment required	Galvanic action can occur within dissimilar metal soldered joints Thorough joint cleaning necessary both before and after soldering No real mechanical strength in soldered joints; strength can only be achieved by combining the soldering with mechanical bonding methods such as crimping – then a permanent joint is produced Maximum working environment temperatures limited to about 150°C Only a severely restricted range of metals can be soldered

Process	Characteristics	Advantages	Disadvantages

Adhesive bonding
Elastomer- and synthetic resin-based compounds
	Joint made by a non-metallic bridging compound
	   Adhesive joining processes do not normally involve heating the joint area, but when heating is required the temperature rarely exceeds 150°C
	   With the exception of thread locking adhesives (anaerobics), bonded joints tend to be permanent
		Dissimilar materials of widely differing thicknesses are easily joined
		   No localized stress concentration areas exist as occurs with spot welds, rivets, bolts, etc.
		   Adhesives tend to be electrical insulators, so electrolytic corrosion in joints of dissimilar material cannot occur
		   Flexible adhesives can absorb shock and vibration
		   Adhesive joining can result in major design simplification and weight reduction
			Bond strength per unit area is limited -especially resistance to peel-type loadings
			   Careful joint preparation is vital
			   Most adhesives have a limited shelf-life
			Few adhesives can be used in environmental conditions above 180°C, and in subzero temperatures joints tend to become brittle
			   Joints are difficult to inspect non-destructively once assembled
			   Elaborate fixturing is sometimes required
			   Many adhesives contain objectionable chemicals or solvents

Mechanical fasteners
Nuts and bolts, screws, rivets, stitching, stapling, retaining clips and rings, locking pins, lock-nuts, press and shrink-fit assemblies
	Methods/devices for positioning and holding together components by inducing residual stresses in the joint	Mechanically fastened joints can be designed for use in a wide variety of environmental conditions
		   Dissimilar materials of widely differing thicknesses and shapes are easily joined
		   Non-permanent joints are readily dismantled and reassembled with simple hand tools
			Localized stresses exist within the joint area, causing mechanical fasteners to often be the site of joint failure
			   Galvanic action can occur within dissimilar metal joints
			   Vibration can cause mechanical and fatigue joint failure
			   Significant joint preparation is often required, e.g. hole drilling and tapping

Bibliography

Bailey, N. et al. 1993. *Welding steels without hydrogen cracking*. Cambridge: Abington Publishing.

British Standards Institution 1984. *Specification for arc welding of carbon and carbon manganese steels – BS 5135*. London: British Standards Institution.

Ciba-Geigy Undated. *User's guide to adhesives*. Duxford, Cambridge: Ciba-Geigy.

Crafer, R. C. & P. J Oakley 1992. *Laser processing in manufacturing*. London: Chapman & Hall.

DeGarmo, E. P., J. T. Black & R. A. Kohser 1988. *Materials and processes in manufacturing*, 7th edn. New York: Macmillan.

Gourd, L. M. 1986. *Principles of welding technology*. London: Edward Arnold.

Leonida, G. 1981. *Handbook of printed circuit design, manufacture, components and assembly*. Ayr: Electrochemical Publications.

Lindberg, R. A. 1983. *Processes and materials of manufacture*. Newton, MA: Allyn & Bacon.

Niebel, B. W., A. B. Draper & R. A. Wysk 1989. *Modern manufacturing process engineering*. New York: McGraw-Hill.

Sanderson, A. 1993. Recent advances in electron beam welding. *Professional Engineering* **6**(5), 20–21.

Technical Editor 1992. Adapting the fastener to the product. *Professional Engineering* **5**(9), 36–37.

Case study

A medium-carbon steel pressure vessel has a number of external piping connections of the form shown in Figs 5.4a and 5.32a. The designer has the following two options of how to attach these connections to the body of the vessel:

    (a) purchase forged steel weldments as illustrated in Figure 5.32b, which can be directly welded onto the pressure vessel

    (b) purchase parallel thick-wall tubing cut to length and plane (flat) flanges, as shown in Figure 5.32c. These would require two welds, one connecting the tubing to the vessel, and the other joining a flange to the tubing.

Consider the pros and cons of each of these alternatives and make recommendations on which option you consider to be the most economical. Bear in mind that this is a thick-wall pressure vessel and, as such, both material and welds are subject to stringent non-destructive testing (NDT), before acceptance by the customer.

Solution

While forgings are expensive, the flange weldment option requires only one (multirun) weld to attach it to the pressure vessel wall. This contrasts with two weld areas necessary when building up the pipe connection from a separate flange and length of thick-walled tubing.

Unfortunately, the solution is not as simple as comparing the cost of a second weld (and its NDT) with the difference in cost between the forged weldment and a plain flange plus length of thick-wall tube. The problem here is one of consistent weld quality, bearing in mind that when welding thick materials multiruns are inevitable.

Because the cylindrical pressure vessel body is the expensive part, and as there is a severe

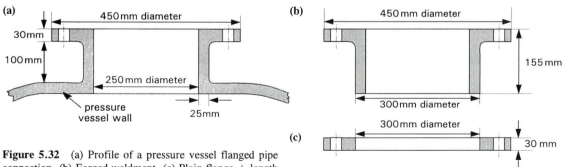

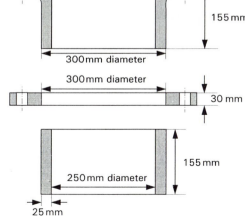

Figure 5.32 (a) Profile of a pressure vessel flanged pipe connection. (b) Forged weldment. (c) Plain flange + length of thick-walled tubing.

limitation on the number of times that defective welds can be gouged out and replaced before unacceptable metallurgical effects in the HAZ occur (§5.2.6), the welding method used must provide reliable weld quality on every weld run performed on the vessel body.

The preferred welding procedure for pressure vessel work, and which offers this consistent weld quality, is submerged arc welding. Unfortunately, in this example the gap between the underside of the weldment flange and the outside of the pressure vessel is only 100 mm, and this is too small to permit anything other than welding by manual methods. But, with manual welding there is always an increased risk of weld defects, so this must be taken into account in assessing the options available.

Considering the two-piece option, if the thick-wall tubing is first welded to the pressure vessel wall, no flange is in position to obstruct the use of sub-arc welding equipment, but when the flange comes to be welded to the other end of the tubing there is again restricted access, making manual welding the only option. However, if there should be quality problems with this manual flange weld, and it has to be replaced, the expensive pressure vessel is not being metallurgically put at risk.

On balance therefore, with access being insufficient to permit sub-arc welding of the one-piece weldment, it could be commercially less risky to use the two-piece option, even though this involves doubling the amount of welding and associated NDT.

Questions

5.1 Retrieve the list of different joint applications that you were asked to produce in the exercise at the end of §5.1. In the light of what you have learned by studying this chapter, re-examine each of the applications on your list and decide whether you still agree with your initial assessment methods.

5.2 Show that you understand the essential differences between the soldering, brazing and fusion welding processes.

5.3 Describe three common types of fusion weld and suggest to which joint configuration each is most suited and why.

5.4 Describe the manual flux-coated electrode arc welding process. What is one of its major limitations and how may it be overcome?

5.5 Why is it necessary to use a flux in some arc welding processes but not in others? What is the largest single disadvantage to the use of flux in the production of multirun welds?

5.6 Sketch and describe suitable joining processes for the following applications, giving reasons for your choices:
   - the joining of overlapping light-gauge stainless-steel sheets used for water-tight containers
   - the joining of low-carbon steel sheet panels to an angular steel frame of heavier section
   - the joining of a ceramic insulating cap over the end of thin-walled metal tubing
   - the joining of an aluminium sheet to a low-carbon steel sheet to be used as a refrigerator heat exchanger panel
   - the joining of two thin, low-carbon steel sheets with an aluminium honeycomb sandwich for use in an aircraft bulkhead – minimum vibration is to be transmitted through the sandwich
   - the joining of thin metal foil;
   - the joining of lengths of thick wall thermoplastic tubing for use as underground cable trunking.

5.7 You are responsible for the design and manufacture of a large pressure vessel destined for use in the petrochemical industry. What major factors would you take into consideration at the design stage and at the manufacturing stage? Which method of welding would you select for the main seam welds, and why?

5.8 Explain the essential differences between arc welding and resistance welding. Give two examples of applications where each would be most appropriate.

5.9 Briefly describe spot, projection and stitch welding, indicating similarities and differences and outlining the relative advantages of each method. How would you spot weld two components which have very different thicknesses?

5.10 What methods of automatic continuous welding may be used for the production of long seam welds in (a) thick medium-carbon steel plate and (b) light-gauge, low-carbon steel sheet.

5.11 Discuss the possibilities of butt welding two cylindrical components of different material composition. Describe briefly one feasible method and state its essential characteristics.

5.12 What are the main similarities between electron beam and laser beam welding? For certain tasks laser beam welding is preferable to electron beam welding. Why?

5.13 Why does the aerospace industry prefer TIG welding to MIG, despite the latter being generally quicker?

5.14 What is meant by the term "weldability of steels"? How is it assessed?

5.15 What are the essential differences between fusion welding and brazing? Suggest at least two applications where brazing would be preferable to welding.

5.16 Discuss the reasons for the increasing popularity of adhesive bonding in engineering. Give typical examples of applications where adhesive bonding has superseded other methods of joining.

5.17 How can the stress distribution of adhesive bonded butt joints be improved?

5.18 All mechanical fasteners are local stress raisers so why are they the most commonly used joining method?

5.19 Determine the costs of making a 60° vee butt weld between two 10 mm thick carbon steel plates 1 m long for each of the three welding methods given in the following table. Assume that all weld preparation machining necessary has been completed.

	Oxy-acetylene	Manual metal arc	Sub-arc
Welding speed (mm/min)	100	250	500
Number of runs required	3	3	1
Cost of filler material (£/kg)	4.50	9.00	7.50
Setting-up time (min)	10	10	5
Joint preparation time (min)	15	15	5
Operating costs (£/h)	12	20	40
Operating efficiency (%)	25	50	85
Energy requirements	3 m³/h of each gas	10 kWh	30 kWh
Power costs	£2.00/m³ average for each gas	£0.28/kWh	£0.28/kWh
Total filler metal needed to complete all weld runs (kg)	1.5	2.0	1.0

Comment upon the comparative information given in the above table and your findings with reference to the welding methods specified.

Explain why the first two methods are still used extensively for the welding of engineering products even though their weld consistency relies much more upon operator skill than is the case with sub-arc welding.

Chapter Six
Surface treatment of materials

After reading this chapter you should understand:
  (a) why surface treatment is important and what the three principal forms of treatment are
  (b) how to choose between mechanical and chemical cleaning and smoothing processes
  (c) the difference between polishing and electropolishing
  (d) why surface protection is so often necessary
  (e) the differences between sacrificial and surface coating protection methods
  (f) what conversion coatings are
  (g) what the electroplating process is and how to design components for efficient plating
  (h) what paints arc and how they are applied
  (i) what vitreous coatings are and where they are used
  (j) how metallic vapours are produced and why they are used as a coating medium
  (k) how to spray molten metal onto a surface to salvage it or to produce a surface offering special properties
  (l) the way that cladded sheet metals are made by diffusion bonding
  (m) which surface coatings can be applied successfully to plastic components.

Any activity carried out on a workpiece involves both time and money and therefore must be economically justified. Surface treatment operations are no exception, so it is essential to ensure that the most appropriate process is employed to achieve the required end result.

Some finishing processes, such as painting, are at times treated as a minor cosmetic activity, adding nothing to the efficient operation of the end product. However, such treatment can be important in enhancing the perceived quality of the product in the eyes of the customer. Therefore costs allocated to surface treatments are a matter of both technical and commercial judgement, and will vary widely from application to application.

Exercise   A vast array of everyday objects are "surface treated". Examine a range of them and try to work out the likely surface treatments employed and their probable cost in relation to total product value.

Component surfaces are treated for three main reasons: to clean them; to improve their surface – usually by smoothing and polishing; to provide them with coatings as a means of surface protection and/or to provide a decorative finish. Each of these areas of application is considered in this chapter, although heat treatments designed solely to increase surface hardness are not included as they are more appropriately and comprehensively dealt with in materials science texts (Smith 1990).

6.1 Surface cleaning methods

Component cleaning involves the removal of unwanted material from the part's surface, and can range from residual sand removal from castings to the removal of a previously applied protective oil film from a highly polished surface. The methods available are equally varied and range from coarse cleaning methods that involve physical removal of heavy surface scale from the part to be cleaned to extremely mild processes involving no more than degreasing and fine polishing.

6.1.1 Mechanical processes

6.1.1.1 Wire brushing

One of the most familiar surface cleaning methods for removing contaminants such as surface scale and rust is the use of a stiff wire brush. It is cheap, easy to use and, while not providing a clean metal finish, is adequate for many non-critical applications. In most cases it is the preliminary step to subsequent operations, the most common of which is painting.

Power-driven rotary brushes are also available and are more suitable than hand brushes for larger jobs.

6.1.1.2 Tumbling

Another coarse cleaning process, tumbling is used to clean and deburr (remove rough edges) castings. The process is simple and involves placing the castings to be cleaned inside a circular steel drum, which is then slowly rotated at typically 10–15 rev/min. This rotary action results in the castings randomly hitting one another sufficiently hard both to remove unwanted sharp edges and to shake free any moulding sand not removed during fettling (§2.4.1.1, step 11)

Figure 6.1   Two-unit rotary vibrator and conveyor. (Courtesy of Rotofinish-Rösler & Co, Prescot.)

Some tumblers include loose material such as sand, granite chips or stones with the castings to enhance the cleaning process, although in this case it is necessary to ensure that all such loose material is removed from the castings when they are unloaded from the tumbler drum.

Tumbling is a cheap and effective way of coarse cleaning, but it is only suitable for castings that are structurally robust. Also, internal surfaces are not satisfactorily cleaned by this method.

6.1.1.3 Barrel and vibratory finishing

Barrel finishing may be considered as a refined form of tumbling (§6.1.1.2), although it also smoothes as well as surface cleans. This is because the non-abrasive material used in tumbling is replaced by abrasive particles in barrel finishing. Artificial abrasives are normally used as their size and texture are easily controlled.

Vibratory motion has taken over from true rotary motion in recent years as it is much more labour-intensive to load and unload barrel-type machines than the easily mechanized rotary vibration machines (Fig. 6.1). Furthermore, vibration frequency can be readily controlled to ensure a more precise and consistent process than is possible with rotary barrel finishers.

6.1.1.4 Jet blast cleaning

One method of avoiding the physical impact between components that occurs in tumbling and barrel finishing is to strike the surface to be cleaned with a suitable fluid, the most commonly used being compressed air containing either steel shot or coarse sand.

Exercise   Why do you think compressed air is not used on its own? It would certainly be much easier and the equipment required would be simpler.
Hint Kinetic energy!

To isolate the jet cleaning operation from the surrounding environment, it is usually carried out in an enclosed cabinet. The component to be cleaned is loaded into the working area of the jet blaster cabinet and the stream of fluid is then directed either manually at the dirty surfaces or, less commonly, the workpiece is moved around under one or more fixed jets. Manual manipulation of either the jet nozzle or component is carried out through a rubber glove box (Fig. 6.2).

The spent shot, sand, etc. falls through a hole in the floor of the working area and is then recirculated.

The compressed air used in jet blasting is at a pressure of up to 6 bar, which is conveniently the typical pressure found in most factory piped air supplies. Much lower pressures are used for non-ferrous parts.

Jet blasting enables the impact fluid to be efficiently directed at internal surfaces, although it is always difficult to clean complex internally cored holes completely or, more importantly, to know whether or not they have been properly cleaned.

An advantage of jet blasting is that the surface finish achieved provides a particularly good key for subsequent painting operations.

Special portable units that invariably use water combined with chemical cleaning agents are frequently used to remove atmospheric contamination from the outside of buildings.

Figure 6.2 Air jet blasting cabinet. (Courtesy of Power Blast Ltd, Camberley.)

6.1.2 Chemical processes

Another highly effective way of removing certain forms of surface contamination is by the use of chemicals.

Safety warning While normal care has to be exercised when using the mechanical equipment described in §6.1.1, the use of chemicals is always a hazardous affair, requiring special care and protection of the personnel engaged in such work. Clearly, the more mechanized this work can become the less the risk to staff. However, there will always be some human intervention and *this calls for special training and care of the staff involved.*

Chemical cleaning is often referred to as degreasing and, as this term implies, most chemical cleaning is solvent based. However, this is not always the case, as is demonstrated by the pickling process (§6.1.2.3).

A wide range of chemical processes are used for surface cleaning, specific ones being recommended for particular surface materials and contaminants to be removed, but this chapter only describes the basic process principles as applied to general engineering applications.

6.1.2.1 Organic solvent cleaning

Oils and greases are the commonest corrosion inhibitors used by industry, but when the parts protected by these substances are required for use it is usually then necessary to remove such protective coatings. Organic solvents are used for this degreasing activity, and they are normally applied by either component immersion, spraying or vapour condensation, the precise method depending upon the size and shape of the parts involved. Application by immersion or spraying is self-explanatory, but precisely how vapour degreasing works requires a brief description.

An organic solvent such as trichloroethylene is heated sufficiently (approximately 100°C) to cause it to vaporize. This vapour is then passed over the component surfaces to be degreased, but because they are cold the solvent condenses on them. The condensate then runs down over the dirty surfaces and, in so doing, dissolves the surface contaminants and washes them away. The dirty solvent is then recirculated, the removed oil and grease being easily separated out when the solvent is revaporized before its next pass through the degreasing cabinet.

It is clear that the key to this whole process is that the surfaces onto which the solvent vapour is required to condense must always be cold, or at least well below the vaporization temperature of the solvent. If this is not so, condensation will not occur and the process does not work. This is generally not a problem with most ferrous components and items of significant mass, but may present difficulties with thin sheet materials, particularly if they have a high thermal conductivity (e.g. aluminium and copper).

6.1.2.2 Alkaline surface cleaning

Organic solvents are an effective cleaning agent as long as the surface contaminants are organic solvent soluble, but if they are not then alkaline solutions are normally used. Furthermore, adding a water-soluble detergent to the alkali creates a highly versatile cleaning medium that will remove organic and most other forms of surface contamination. This explains why alkali-based cleaning is the most popular of all chemical cleaning processes.

Alkaline solutions can be applied to the work surfaces by immersion or spraying, and are also used as active agents in steam cleaning when the parts to be cleaned are too large to be dipped. To ensure that all traces of alkaline solution are removed after cleaning it is essential to rinse components thoroughly with water after this form of cleaning.

Steel and cast iron parts are particularly suitable for this type of cleaning. However, because of steel's propensity for rusting, thorough drying is essential if the parts are not used immediately after the water rinse.

Certain non-ferrous metals such as aluminium, zinc, tin and brass should not be alkali cleaned.

6.1.2.3 Pickling

Pickling may be defined as a process that removes oxide layers from a metal surface using acids, and which may or may not result in erosion of the base metal (Roessell & Swain 1990). Thus, pickling is more a chemical descaling process than a surface cleaning one, although it will be partially ineffective if, before pickling, the component's surface is not properly degreased.

Figure 6.3 Stainless-steel fabrication before and after pickling. (Courtesy of Anopol Ltd, Birmingham.)

Its principal uses are in the removal of mill scale from hot rolled steel and removal of surface scale formed during certain types of heat treatment process. It is also used for removing weld scale and heat discoloration from sheet metal fabrications (Fig. 6.3).

It is essential that overpickling does not occur (i.e. leaving the component in the pickling bath for too long), as this results in surface pitting and, in the case of steels, hydrogen embrittlement problems.

After pickling all traces of acid must be removed, and this is normally achieved by first dipping in an alkaline solution, followed by a thorough clean water rinse.

6.1.3 Ultrasonic cleaning

This cleaning process is very different from those so far described, and an understanding of its operating principle is essential if its many potential uses are to be fully appreciated.

The operating principle is as follows. If a low-viscosity liquid such as water is subjected to high-frequency (25–40 kHz) ultrasonic wave energy, or ultrasound as it is called, intense cavitation is generated within the fluid. Cavitation is the generation and growth of vast quantities of gaseous bubbles within the fluid which, upon reaching a certain size, typically 0.15 mm diameter, implode (collapse inwards). As each bubble collapses (Fig. 6.4), a microscopic inward moving jet of fluid is created at an astonishing 250 mph, and it is this jet that blasts away the contamination from the dirty surface.

The sustained generation, growth and implosion of huge numbers of bubbles throughout the ultrasonic bath, and on all surfaces of the submerged component to be cleaned, is referred to as the cavitation effect.

Commercial ultrasonic cleaning units usually consist of a bath of fluid specially selected to suit both the material to be cleaned and the type of contaminant to be removed, plus an electri-

Figure 6.4 0.15 mm diameter cavitation bubble during implosion. (Courtesy of Professor L. A. Crum, University of Washington, Seattle, USA.)

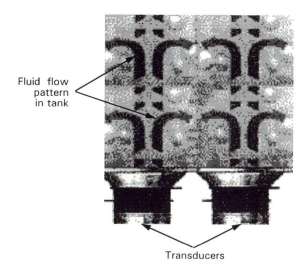

Fluid flow
pattern
in tank

Transducers

Figure 6.5 Ultrasonic transducers attached to tank bottom. (Courtesy of Langford Ultrasonics Ltd, Birmingham.)

cal transducer unit for generating the required ultrasound. The transducer(s) may be attached to the bath in a number of ways but is frequently externally bonded to the bottom of the fluid tank (Fig. 6.5).

Portable units can also be obtained for fitting into existing tanks, transducer power required being approximately 10 W per litre of tank capacity.

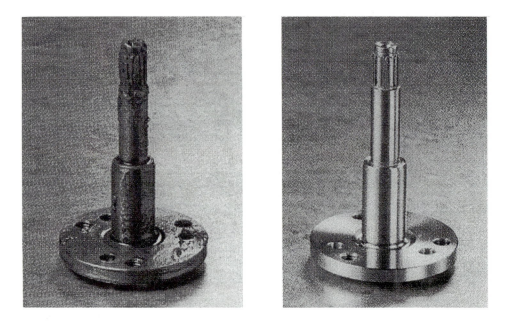

Figure 6.6 Before and after ultrasonic cleaning (cleaning time 3 minutes). (Courtesy of Langford Ultrasonics Ltd, Birmingham.)

Cleaning times are typically 2–5 minutes depending upon the level of contamination and component geometric complexity (Fig. 6.6). Cleaning speeds are generally quicker and more thorough than is achievable with most other methods, and both intricate and delicate components can be cleaned without risk of damage.

Exercise When you next go to an optician ask whether an ultrasonic bath is used for cleaning spectacles; this is usually the case. If you are in luck ask to see your old spectacles or sunglasses being cleaned. You will be amazed how much dirt is removed and to find that it is dislodged from *all* areas of the frame.

6.2 Surface smoothing processes

6.2.1 Hand grinding

This is often erroneously classed as a cleaning process despite the fact that significant metal removal is usually involved. It is more a fettling/deburring/smoothing activity. Many surfaces need dressing or smoothing, an excellent example being weld dressing (Fig. 6.7).

Because the abrasive head used is mounted in a hand-held portable grinder (Fig. 6.7), the method offers great flexibility but is labour intensive.

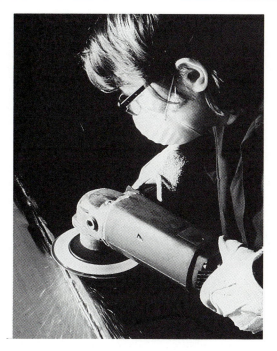

Figure 6.7 Weld dressing by hand grinding. (Courtesy of 3M United Kingdom Ltd, Bracknell.)

6.2.2 Belt sanding

Like hand grinding, this is also an abrasive process, the principal difference being that, instead of the abrasive being presented to the workpiece in the form of a rotating wheel or disk, the abrasive grits are glued to one side of a flat strip of flexible backing material (like sandpaper and emery paper). A suitable length of this abrasive strip is made up to form an endless loop, which is then slipped over the drive rollers of the belt sander.

Belt sanders come in both fixed and hand-held formats (Fig. 6.8a and b), the former being only really suitable for flat surfaces. Crevices, corners, and internal areas are usually inaccessible to belt sanders.

The surface finish produced comprises a mass of fine scratches, their fineness being determined by the grit size of the abrasive belt used.

6.2.3 Polishing and buffing

Occasionally it is necessary to enhance the appearance of an already clean and smooth surface still further by polishing. Like the previous two processes described, this too is an abrasive operation, although the abrasive used is extremely fine, and when applied to the workpiece surface it cuts just sufficiently to remove the multiplicity of fine scratches that constitute most matt finish surfaces (Fig. 6.9). The end result should be a highly polished surface devoid of visible imperfections.

Polishing is performed by applying the fine abrasive to a smooth, lint-free cloth and then rubbing it on the work surface by hand in a rotary or reciprocating action. Alternatively, a cloth polishing wheel can be charged with abrasive and the workpiece to be polished passed over the wheel's surface until the required surface finish is achieved (Fig. 6.10).

(a)

Figure 6.8 (a) Fixed belt sander. (b) Hand-held belt sander. (Courtesy of 3M United Kingdom Ltd, Bracknell.)

(b)

Figure 6.9 *Highly magnified* surface before polishing.

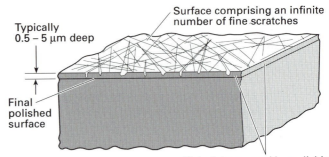

Typically 0.5 – 5 μm deep

Surface comprising an infinite number of fine scratches

Final polished surface

Material removed by polishing to eliminate scratches

Figure 6.10 Polishing a turbine blade. (Courtesy of 3M United Kingdom Ltd, Bracknell.)

Buffing is an extension to polishing using an even finer abrasive such as jeweller's rouge. It is used after polishing when an exceptionally smooth surface finish is required. Mirror finishes are then possible, but both polishing and buffing are labour-intensive processes, and thus expensive, and should only be used when absolutely necessary for technical and/or commercial reasons.

6.2.4 Electropolishing

Electropolishing is basically the reverse of electroplating (§6.3.2.2) in that a very thin layer (up to 40 μm) of the component's surface is stripped away by deplating. This is achieved by reversing the polarity of a normal electroplating cell (Fig. 6.13) in the following way.

The component is made the anode (+), and placed adjacent to carefully designed and shaped cathode (–) plates. Both are immersed in a tank filled with a suitable electrolyte such as phosphoric acid. When a high anodic current is applied, the surface layer of the workpiece migrates towards the cathode plates. Fortunately, this deplating action is not completely uniform across the whole job profile, there being a preferential dissolving of the peaks of the surface profile. This helps to produce a smooth bright finish (Anopol 1990).

Electropolishing is mainly a finishing operation for machined (usually ground) surfaces and is particularly suited to awkward shapes that do not lend themselves to polishing or buffing. The final polished surface is also stress free, which is ideal for such applications as surgical instruments and the inside of watch cases. It has also been shown that bacteriological growth occurs much more slowly on electropolished surfaces than on mechanically finished ones (Anopol 1992), thus making this process of particular benefit to the pharmaceutical, food and brewing industries (Fig. 6.11).

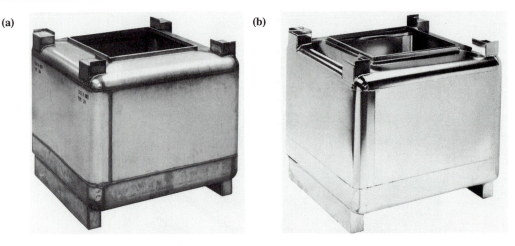

Figure 6.11 (a) Pharmaceutical container before electropolishing. (b) Pharmaceutical container after electropolishing. (Courtesy of Anopol Ltd, Birmingham.)

6.3 Surface protection

The surface of a component frequently needs to be protected from its immediate environment or enhanced to resist wear under normal operating conditions. Selecting the correct treatment and applying it correctly can make the difference between a well-made product and a shoddy, unreliable, short-lived one.

The most familiar surface protection methods involve either applying a coating of material that is resistant to environmental attack or that changes the surface layer chemically to achieve the same effect. These are called direct protection processes. There is, however, another approach that relies upon a completely different principle to resist surface attack, and this is termed sacrificial protection.

Exercise List a range of everyday items that have some form of surface protection layer applied to them. Can you see what coatings have been used and what hostile elements they are designed to resist? Put your list aside for use later (Question 6.1).

6.3.1 Sacrificial protection

As its name suggests, this process relies upon sacrificing one material to protect another (the workpiece).

A metal surrounded by a hostile environment, for example low-carbon steel in normal atmosphere or submerged in seawater, will corrode (oxidize) extremely quickly. However, if lumps of certain other metals, such as zinc, cadmium or aluminium, are attached to it, the attacking medium will, surprisingly, largely leave the steel surface alone and attack only the sacrificial metal, even although areas of the steel are fully exposed to the hostile atmosphere. This will continue until all the sacrificial metal has been eroded, after which all such *cathodic protection* is lost and the steel will start to corrode if a new supply of sacrificial metal is not provided.

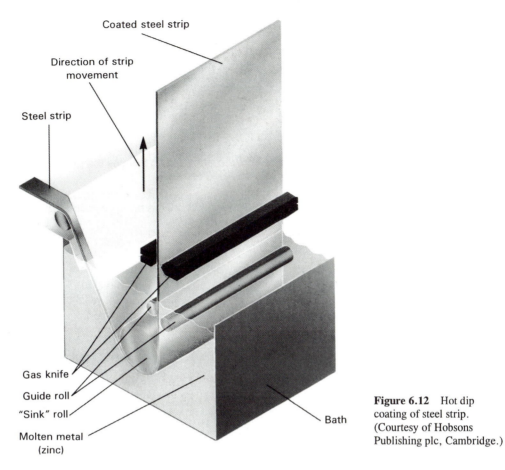

Coated steel strip

Direction of strip movement

Steel strip

Gas knife
Guide roll
"Sink" roll
Molten metal
(zinc)

Bath

Figure 6.12 Hot dip coating of steel strip. (Courtesy of Hobsons Publishing plc, Cambridge.)

In addition to the application of corrosion-resistant and antifouling paint, this is a cheap and popular method of protection for ships' steel hulls. Pieces of zinc are placed at strategic points along the hull's length below the water line. They are replaced periodically to maintain corrosion resistance.

A more common example of sacrificial protection is the protection of steel sheet by coating it with zinc. This is called galvanized sheet and is commonly used as a cheap roofing material and as a steel coating in vehicle body panel manufacture. It is normally applied either by electroplating (§6.3.2.2) or by hot dip coating, in which the steel is passed through a bath of molten zinc (Fig. 6.12).

Table 6.1 (Goodman & Fewtrell 1992) illustrates how common hot-dip coating of steel has become in recent years, and indicates typical coating thicknesses used.

6.3.2 Direct protection methods

6.3.2.1 Chemical conversion coatings
These protective coatings are not produced by applying a coating material over the component's surface but are generated by chemically changing its surface into one that provides the desired

Table 6.1 Hot dip sacrificial coatings for steel sheet.

Metal coating	Coating thickness (μm)	Typical applications
Zinc	7–45	Building materials
Iron/zinc alloy	4–14	Automotive body panels
95% zinc/5% aluminium	7–25	Building materials
45% zinc/55% aluminium	20–25	Automotive body panels and automotive exhausts
Aluminium	7–25	Petrochemical installations and high-temperature applications

properties. These properties range from enhanced corrosion resistance, and an improved surface for subsequent painting, to lubricant retention on sheet, wire and tube stock to enhance their cold working properties.

The three principal types of conversion coating are phosphate, chromate and anodic, the first two being applied to both ferrous and non-ferrous metals, with anodic coatings being mainly suitable for aluminium and magnesium and their alloys.

Phosphating is most widely used as a form of metal pretreatment (Freeman 1986), particularly as a preparation for steel before painting. A dilute solution of phosphoric acid is applied to the component's surface – usually iron, steel, aluminium or zinc. This sets up a chemical reaction that changes the surface into a corrosion-resistant phosphate layer to a depth of up to 75 μm. *As with all conversion coatings, no significant increase in component dimensions occurs.*

Note Do-it-yourself preparations to prevent further rusting of vehicle bodies often work on the conversion coating principle. Loose surface rust is removed and a phosphate coating is then painted on to convert chemically any remaining embedded rust, and the bare metal surfaces, into a rust-resistant layer. When dry this layer can then be painted over with normal "touch-up" body paint.

Chromating is similar to phosphating in that it also chemically creates a surface that provides both a good base for subsequent painting and enhanced corrosion resistance. Chromium is the base metal of the acid solutions used, but the thickness of the conversion layer is very small – only of the order of 0.01 μm. While chromating can be applied to ferrous metals, it is particularly suited to aluminium, magnesium, zinc, cadmium and their alloys.

Anodizing, although still a conversion coating process, is achieved in a different way to phosphating and chromating. Instead of applying an acid solution by dipping or spraying, the component is wired up as the anode of an electrolytic cell and immersed in an acid bath. The two metals that are normally anodized are aluminium and magnesium, and the effect of their being immersed in this electrolytic bath is to increase greatly the 1–2 μm oxide layer that forms naturally in air on these two metals.

Although anodizing is used to prepare surfaces for painting, anodized surfaces are extensively employed in the aerospace industry in their unpainted condition. However, unpainted surfaces are usually sealed by immersion in hot water. The surface can also be dyed after anodizing to give an enhanced visual appearance if required.

Because components do not change size significantly during anodizing, if used in the unpainted state finish machined precision parts can be anodized without making any allowance for surface coating thickness.

6.3.2.2 Electroplating

Second only to painting, electroplated surfaces are probably the most familiar of all surface finishes. Chromium plating used to reign supreme in the automotive industry, although because of its high cost it is now only to be seen to any degree on a few luxury vehicles. However, it is still found on many domestic products and office furniture.

Nevertheless, electroplating is still an extremely important surface finish, and, providing the correct preplating treatment is carried out, its flexibility and attractive appearance make it a process worthy of consideration in the right circumstances.

All plating processes operate on the same electrochemical principle (Fig. 6.13).

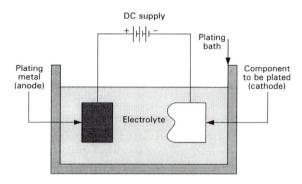

Figure 6.13 Basic electroplating cell.

Table 6.2 Metals commonly used for plating

Plating metal	Coating thicknesses (μm)	Characteristics/ principal uses
Zinc	2.5–25	Preferred alternative to galvanizing of small iron and steel parts as the coating is more even than with dipping. Used extensively in domestic appliance and automotive body panels
Cadmium	5–10	Competes with zinc, although much more expensive. Thus it is only used when its superior properties such as solderability and lower electrical contact resistance are important, e.g. in the electronics industry; also superior to zinc in marine environments
Tin and its alloys	0.1–2	Other method of tin coating of steel strip. Coating more even than dip coating so favoured for sheet steel used in production of food containers, particularly as it is not toxic
Precious metals (e.g. silver, gold, platinum)	0.5–100	Owing to the very high cost of these plating metals, they used to be confined to the jewellery industry. However, they are now used extensively in the electronics industry because of their extremely low contact resistance
Copper and its alloys	10–50	Copper, bronze and brass are all used extensively for domestic articles, brass being usually sealed with a layer of clear lacquer. Copper is used as the base layer for chrome plating, and as a means of increasing surface heat transfer properties. Brass plating is good as a base layer for bonding rubber to metal
Nickel	8–500	An intermediate layer between copper and chromium. Offers excellent wear and corrosion resistance but is expensive. Should not be applied to ferrous surfaces without a primary layer of copper
Chromium (decorative)	0.3–0.8	Purely a decorative finish, and for best results should be underlaid by layers of copper and nickel. Used mainly for domestic products and office furniture. No longer used much by the automotive industry as it has been superseded by aluminium and stainless steel because of the long processing time and cost constraints of the plating process
Chromium (hard industrial)	75–500	Used as a hard, wear-resistant surface. It is applied directly to such ferrous parts as moulding die surfaces, internal combustion engine piston rings and exhaust systems. Also used to reclaim inspection plug gauges (§11.5.3), the chromed surface being ground to size after plating

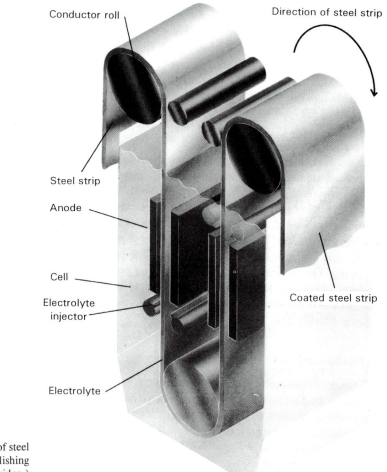

Conductor roll

Direction of steel strip

Steel strip

Anode

Cell

Electrolyte
injector

Coated steel strip

Electrolyte

Figure 6.14 Electroplating of steel
strip. (Courtesy of Hobsons Publishing
plc, Cambridge.)

The component to be plated is wired into a DC circuit as the cathode (−) with the metal used as
the plating source as the anode (+). Both anode and cathode are immersed in a bath of suitable
electrolyte containing salts of the metal to be plated. When the circuit is switched on, metal from
the electrolyte is deposited on the workpiece and is continually replenished from the anode.

 Most metals can be electroplated using this process, but it is essential that components to be
plated are designed to make the end result as attractive as possible at the lowest cost. Rarely is
a part impossible to plate, but its design might make the process unnecessarily difficult.

Electroplating can also be operated as a continuous process, and is used in this way when
producing plated continuous strip (Fig. 6.14).

The two greatest problems with plating are (1) the metal is never deposited uniformly, with
deep recesses presenting the biggest difficulty; and (2) if the part does not have all its surface
imperfections polished out before plating these marks will be perpetuated during plating and so
mar the look of the finished surface.

Figure 6.15 (Committee for the Promotion of Electroplating 1983) offers a pictorial guide to
good design of components that are to be plated.

The most commonly plated metals and their principal uses are listed in Table 6.2.

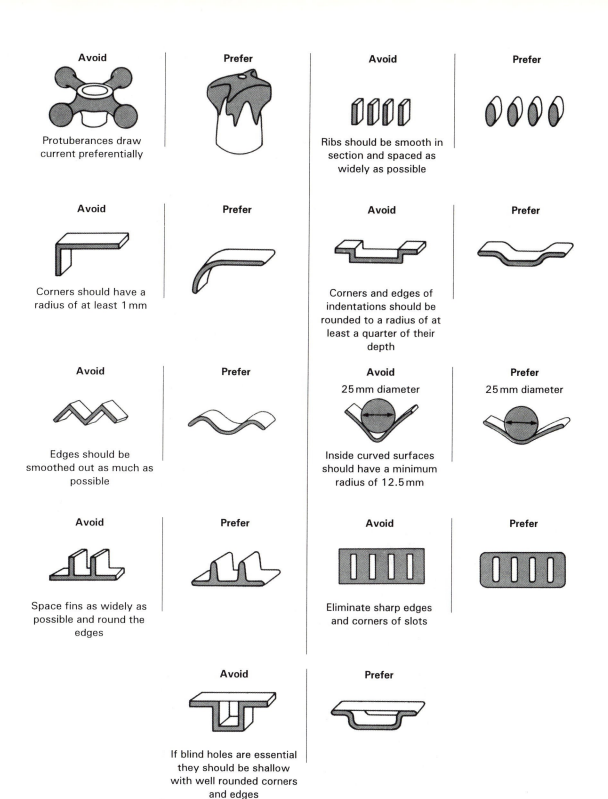

Avoid

Protuberances draw current preferentially

Prefer

Avoid

Ribs should be smooth in section and spaced as widely as possible

Prefer

Avoid

Corners should have a radius of at least 1 mm

Prefer

Avoid

Corners and edges of indentations should be rounded to a radius of at least a quarter of their depth

Prefer

Avoid

Edges should be smoothed out as much as possible

Prefer

Avoid
25 mm diameter

Inside curved surfaces should have a minimum radius of 12.5 mm

Prefer
25 mm diameter

Avoid

Space fins as widely as possible and round the edges

Prefer

Avoid

Eliminate sharp edges and corners of slots

Prefer

Avoid

If blind holes are essential they should be shallow with well rounded corners and edges

Prefer

Figure 6.15 Good and not so good design features of plated components. (Courtesy of Metal Finishing Association, Birmingham.)

6.3.2.3 Organic coatings – paints

Painting is the most common form of surface coating and is found in all domestic and commercial environments. Its popularity is undoubtedly due to the wide range of paints available to meet differing applications, the ease with which they can generally be applied and the attractive finish produced, providing the surface is adequately prepared and the paint correctly applied.

Paint is the term used to cover organic (natural and synthetic) surface coatings, and includes both oil- and polymer- (plastic-) based paints, enamels and lacquers.

The basic constituents of paint are (Lambourne 1987):

(a) the material that forms the continuous film on the surface to be covered

(b) pigment(s) to provide the opacity (covering power), colour and surface texture (matt, gloss, etc.) required

(c) a solvent or carrier that enables the paint to be easily applied.

In clear enamel, varnish and lacquer finishes the colour pigment is, of course, omitted.

Materials used to create the surface film are made from naturally occurring or synthetic resins and from polymers (plastics) and influence factors such as the paint's covering power and colour intensity.

Oil-based paints typical of those used domestically are not suitable for the bulk of engineering applications, partly because of their long drying time.

Industrial applications usually require coatings that are not only durable and attractive, but are easy and quick to apply. They are also frequently required to offer a significant degree of surface hardness. These are the main reasons why most oil-based paints have now been superseded by either water- or polymer-based ones, many of the thermoplastics and thermosets discussed in Chapter 4 being used for various applications.

A solvent or carrier is only present to enable paint to be made and applied, and plays no long-term role in the paint's performance (Lambourne 1987). After the paint has been applied the solvent evaporates. Thermosetting plastic-based paints harden as a result of polymerization (§4.1.2). This process can be accelerated by the application of heat, and is common practice in the automotive industry, where drying times must be as short as possible.

For any painting operation to be successful thorough surface preparation is essential. If the surface is not free of all contamination (scale, rust, oil, grease, etc.) and well prepared, reliable bonding will not be achieved.

In engineering metals are by far the most common material to be painted, and one of the best surface preparations is a phosphate conversion coating layer (§6.3.2.1). This offers a degree of corrosion resistance in its own right as well providing a slightly roughened surface that is ideal for good paint bonding.

A durable surface can rarely be produced with just one coat, the minimum usually being a primer followed by a finishing coat. In practice, especially with metals, unless the substrate (the surface to be coated) has already been pretreated, three or more layers are normally required to give the long-term durability and quality of finish necessary (Fig. 6.16).

Precoating of steel strip and sheet is an increasingly important activity for major steel producers because of the vast quantities used by the automotive and domestic appliance industries. This is achieved by a continuous process called roller coating (Fig. 6.17) and is similar in principle to painting ceilings and walls using a hand-held roller. Both single- and double-sided coatings are available.

Other methods of applying paint are spreading using a brush, spraying, dip and flow coating and electrodeposition. Spraying is the most commonly used industrial painting method, the spray nozzle being either manually manipulated (Fig. 6.18) or moved through a preprogrammed cycle by a robotic arm.

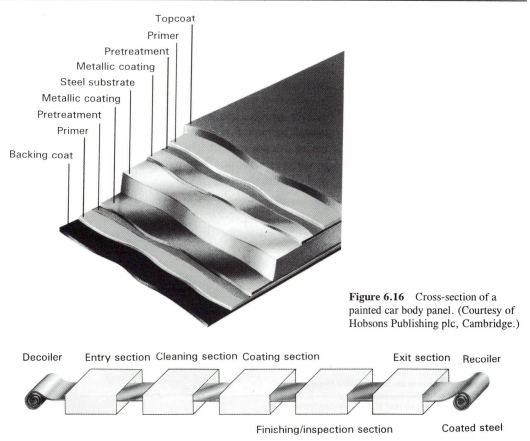

Figure 6.16 Cross-section of a painted car body panel. (Courtesy of Hobsons Publishing plc, Cambridge.)

Decoiler   Entry section   Cleaning section   Coating section   Exit section   Recoiler

Finishing/inspection section   Coated steel

Figure 6.17 Roller coating of steel strip. (Courtesy of Hobsons Publishing plc, Cambridge.)

Figure 6.18 Manual paint spraying. (Courtesy of Kremlin Spray Painting Equipment Ltd, Slough.)

Safety warning Because paints and their solvents are often both toxic and a fire hazard, much care and attention is needed to protect operatives by providing specialist clothing and efficient air extraction systems in the spray booth area.

Not every paint can be sprayed. so it is essential that the correct paint is specified for the application method used.

A problem with spray painting is that, in producing a normal 25–50 μm thick uniform coating, over half of the paint sprayed may unavoidably miss the job and be wasted. Paint loss can be reduced to only 10 per cent if electrospraying is employed. However, special equipment and electrically conductive paint are required as this process involves electrostatically charging the atomized paint as it leaves the spray nozzle so that it is attracted to the oppositely charged workpiece.

Flow coating is particularly useful for painting large complex areas. The workpiece is contained in a booth and paint is forced under pressure through a number of jets that are carefully positioned to ensure complete surface coverage. Excess paint drains off both surface and surrounding booth walls and can be re-used. Paint penetration into recesses is better than can be achieved by simply dipping the part into a paint-filled vat (dip coating) as, despite the simplicity of dip coating, there is frequently a problem of incomplete coverage due to air pockets in component recesses.

Electrodeposition, or electrocoating as it is called in many texts, has gained much favour in the automotive industry for its ability to provide extremely uniform coatings with no paint runs. Although its operating principle is similar to electrospraying in that the workpiece and paint are electrically charged with opposite polarities, the main differences are that in this case a water-based paint is used and the workpiece is immersed in the paint rather than the paint being sprayed on. As the layer of paint becomes thicker the electrical resistance eventually becomes too high for further deposition; this enables coating thickness – typically 20–40 μm – to be easily controlled by varying the electrical settings used.

The automotive industry uses this method mainly for steel priming coats, and has virtually abandoned all other methods for this particular application.

6.3.2.4 Powder spraying

An increasingly popular surface coating process involves the use plastic powders as the covering medium. The powder is usually applied by electrostatic spraying in a similar manner to the electrospraying of paint (§6.3.2.3), although there are differences between the two processes, the most significant being that the powder is electrically charged (up to 90 kV) within the spray gun rather than after it leaves the spray unit, as is the case with electrostatic paint spraying. This has the advantage of confining high voltages to the spray gun (Fig. 6.19).

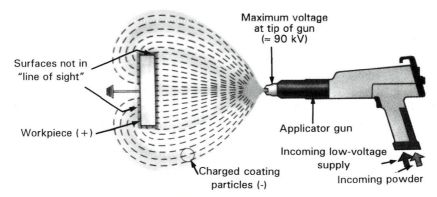

Figure 6.19 Electrostatic powder spraying principle. (Courtesy of ITW Ransburg Gema, Bournemouth.)

The powders used are either thermoplastics such as nylon, PVC and PTFE or thermosetting plastics such as epoxy resins and polyesters, and can be used to coat most metal surfaces.

Because of the electrostatic attraction principle involved, surfaces that are not in the line of sight of the applicator gun can still be coated with ease (Fig. 6.19).

Unlike paints, powders do not contain solvents, and this therefore makes powder spraying much less hazardous and more environmentally friendly than paint spraying processes. As powder-sprayed surfaces do not suffer from paint runs, thicker coatings can be applied and surface blemishes are easily repaired.

Spray units can be either hand held or incorporated into fully automated production lines.

6.3.2.5 Vitreous coatings

These protective coatings use ceramics, porcelain and enamel as the surface sealing medium. While they can be applied to aluminium, their main applications are for coating iron and steel components subject to high temperature, and they offer a glass-like surface with excellent chemical and abrasion resistance. They are frequently used as coatings for cooking utensils, but unfortunately the coatings tend to be brittle and are therefore prone to surface chipping.

Numerous methods of application are possible, but the most common is by applying the coating in either granular form or as a water-based slurry. Heating to approximately $800\,^{\circ}$C for a few minutes produces a visually attractive glass-like surface. Typical coating thickness is in the range of 75–$100\,\mu$m.

6.3.2.6 Vapour-deposited metal coatings

There are situations when only an extremely thin film (less than $0.1\,\mu$m) of metal is required, but to achieve this uniformly over a complete surface is very difficult. However, vapour deposition offers a means of achieving this objective, and at an economic price if carried out on a commercial scale.

The basic process involves placing the workpiece and the coating metal into a vacuum chamber and then heating the coating metal (the evaporant) to its vaporization temperature. The metal vapour then travels outwards in line of sight from its source and is deposited onto all surrounding cold surfaces, including the substrate (workpiece) (Fig. 6.20).

Suitable fixturing design and masking of areas that are not to be coated minimizes the waste of coating metal (Mattox 1993) – particularly important when it is gold or platinum!

The vapour deposition principle is not complex, but its execution is often difficult, the main problem being to achieve satisfactory metal vaporization. This is because many metals do not vaporize until temperatures way above their melting point are reached (White 1968), and some that do still require a very high vaporization temperature, e.g. titanium's vaporization temperature is $3200\,^{\circ}$C.

While heating the evaporant is one way of generating the required metal vapour, in recent years the process of magnetron sputtering has become the main evaporating technique because it does not require the evaporant to be heated. Sputtering is the process of bombarding the evaporant's surface with fast-moving ions (electrically charged atoms) of an inert gas – usually argon. These ions act like armour-piercing bullets, and when they hit the evaporant they knock out atoms from its surface, which then diffuse away as a vapour in the direction of the substrate. When they come into contact with the workpiece they adhere strongly to its surface and so form the required coating, typically 0.05–$0.1\,\mu$m thick. Virtually any metal can be sputter deposited without subjecting the workpiece to a high-temperature environment.

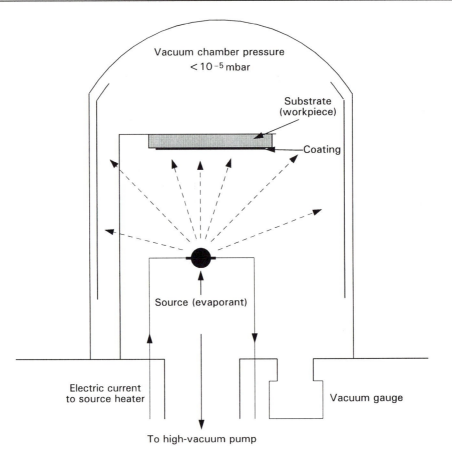

Figure 6.20 Vacuum coating by thermal evaporation from a heated source. (Courtesy of Edwards High Vacuum International, Crawley.)

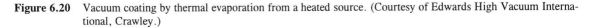

Note A magnetron is a form of electrical valve that, when subjected to a strong magnetic field, generates a high-power, high-frequency energy source. In addition to being the energy source for sputtering, magnetrons are also used in such diverse fields as domestic microwave ovens and in radar applications.

Thermal vapour deposition has been used for many years to coat items with extremely thin layers of precious metals, particularly in the jewellery and allied industries. However, with the introduction of magnetron sputtering the process has become particularly popular in the food packaging industry. For example, the insides of most plastic crisp packets are now coated with a thin layer of aluminium to prevent the atmospheric ingress that occurs if uncoated plastic bags are employed as a result of the porous nature of the plastic normally used.

Most cutting tools (§8.7.1) are now vapour deposition coated with thin films of such metals as titanium nitride, aluminium oxide and titanium carbide to enhance their performance and cutting life. In this application area coatings of up to $10\,\mu$m are applied.

6.3.2.7 Metal spraying (metallizing)

Unfortunately, there are occasions when a component is made undersize, and unless mating parts can be modified this will result in the part being scrapped. However, it is sometimes possible to retrieve this situation by building up the undersize part by metal spraying, also known as metallizing.

The process of metallizing involves firing a stream of fine molten metal particles at the surface of the workpiece which, upon impact, then mechanically bond themselves securely to the surface. When solidified they form the required new metal layer. The more metal that is sprayed onto the substrate the thicker the coating that is built up.

Because metal-sprayed surfaces can be produced from most metals, in addition to being a reclamation/repair process, metal spraying is also used extensively to produce special coatings offering specific mechanical properties such as surface hardness (a form of hard facing), corrosion resistance, wear resistance and resistance to high-temperature working environments.

There are three basic methods of creating and delivering the necessary molten metal spray, and these are flame spraying, arc spraying and plasma spraying.

(a) *Flame spraying* (Fig. 6.21). The spray nozzle is supplied with a constant supply of the metal to be applied, in either wire rod or powder form. As it leaves the nozzle the metal is melted in an oxy-fuel flame and the resulting molten metal is then atomized and projected onto the workpiece by compressed air. Finally, the spray solidifies forming a dense, strongly adherent coating.

The powder supply option enables metals that are not suitable for making into wire form to still be applied without difficulty.

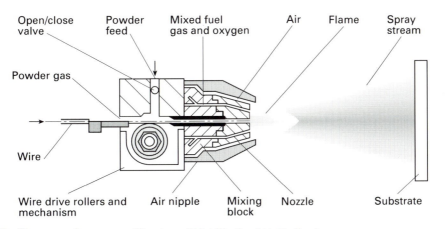

Figure 6.21   Flame spraying process. (Courtesy of Metallisation Ltd, Dudley.)

(b) *Arc spraying* (Fig. 6.22). The spray nozzle is fed with a pair of wires made from the coating metal to be used. They are brought into sufficiently close proximity at the exit end of the nozzle that, when supplied with a high-current, low-voltage DC power supply, they sustain an electric arc. The temperature generated in the arc is in excess of 4000 °C, which is more than sufficient to melt any metal likely to be required. As with flame spraying, the molten metal is then atomized and projected onto the workpiece by compressed air. The coating produced has a greater bond strength than that achieved by flame spraying, and it does not require an oxy-gas supply – only an electrical source and a compressed air supply.

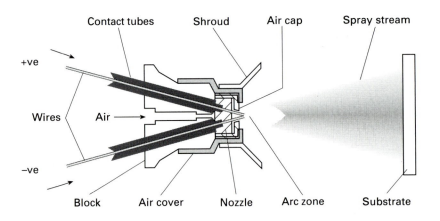

Figure 6.22   Arc spraying process. (Courtesy of Metallisation Ltd, Dudley.)

Note   Can you see the similarity between this process and arc welding? However, there are some important differences. If you are not sure what these are reread Chapter 5, §5.2.1.

(c) *Plasma spraying* (Fig. 6.23). Here an ionized plasma arc is established using a plasma gas mixture of either argon and hydrogen or nitrogen and hydrogen. As with other plasma arc applications (§5.2.1.4), an extremely high temperature (in excess of $15\,000\,°C$) is generated as the plasma jet emerges from the nozzle. While this permits vaporizing of any metal it can also spray ceramics and other materials with high vaporiz-

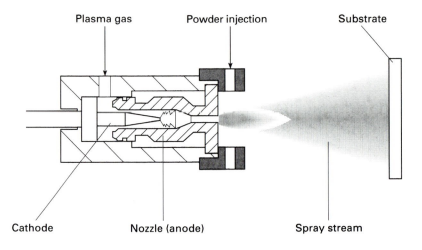

Figure 6.23   Plasma spraying process. (Courtesy of Metallisation Ltd, Dudley.)

ing temperatures. No compressed air is required to atomize the powder material supply, as it is already in a vaporized state when it leaves the plasma arc region, and is somewhat akin to conditions encountered in vapour deposition coating (§6.3.2.6). Also, this minimizes the temperature rise at the surface of the workpiece, while providing a coating of exceptional uniformity and metallurgical integrity.

Whichever spraying method is used, there must be relative motion between nozzle and workpiece if a uniform coating is to be produced. This can be achieved either by hand, as most systems are available as portable units, or by rigidly mounting the nozzle in a machine tool (such as a centre lathe) and traversing along the component as it slowly rotates.

Theoretically thermal coating spray thicknesses have no limit, but in practice they range from $50\,\mu m$ to about $10\,mm$.

As with all coating methods correct surface preparation is essential. In the case of metallizing a rough surface is preferred, usually achieved by jet blasting (§6.1.1.4), as such surfaces provide an excellent base onto which the molten metal spray can adhere.

Most spray-coated surfaces tend to be slightly porous, with coatings reaching approximately 85–90 per cent of wrought metal density. This can be used to advantage as the surface porosity can be used to retain lubricants, so making it an attractive finish for bearing surfaces.

Unfortunately, noise levels generated can be high, particularly when flame-spraying rod format metals, and can reach $110\,dB(A)$. Therefore operator ear protection is essential.

6.3.2.8 Cladding (diffusion bonding)

It is becoming increasingly common, particularly in the automotive and aerospace industries, to manufacture parts from sheet material that has already been surface coated (§6.3.2.3). Paint is by no means the only precoating material used, and metal (and plastics) are now also being used extensively.

The base material – frequently low-carbon steel – can be clad on either one or both sides with coating metal by pressing the sandwich together under high pressure and at a temperature approximately half of the melting point of the base metal for a suitable length of time, typically 1 hour. This combination of pressure, temperature and time results in a condition being established at the sandwich interface(s) known as diffusion bonding.

Note Diffusion is the movement and mixing of atoms from one material to another, and is the reason why it is possible to dissolve solids in liquids.

When sufficient transfer of atoms at the interface has occurred – hence the time factor in the process – a strong and permanent bond will have been created. Providing adequate pressure and a sufficiently slow traversing rate are used, it is possible to diffusion bond strip as a continuous process by high-pressure rolling. However, whatever method is used, as with all other coating processes, surface preparation in the form of thorough cleaning and degreasing is vital if bonding is to be successful.

Typical metallic bonded combinations are the cladding of carbon steel with titanium, stainless steel, zinc, nickel, aluminium and copper, and the cladding of aluminium with harder non-ferrous and ferrous metals.

When working with clad materials in which the cladding is on only one side, special care is necessary to ensure that the cladding material is facing the correct way in service. Also, special treatment may be necessary to ensure that component cut edges do not present a source of subsequent surface failure.

6.4 Coating plastics

The majority of plastic components do not require any surface coating as most plastics give an aesthetically pleasing appearance anyway. However, there are occasions when a special coating is necessary, possibly to provide better abrasion resistance or a metallic appearance or to act as a surface sealant.

While metallizing can be used (§6.3.2.7), the most common coating process for plastics is electroplating (§6.3.2.2). The electroplating process requires that the material to be plated be electrically conductive, but unfortunately most plastics are not. Therefore to plate plastics successfully it is first necessary to chemically etch or sand blast the surface to make it rough. The roughened surface is then dipped into a copper or other metallic-based liquid to provide the electrically conductive surface needed for the electroplating process to work.

Bright chromium and nickel are the most widely used plating materials, particularly by the automotive industry and by manufacturers of domestic fittings (Fig. 6.24).

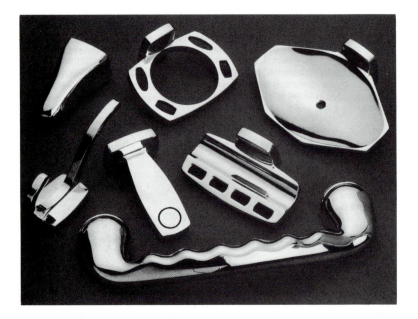

Figure 6.24   Typical range of chrome-plated plastic bathroom fittings. (Courtesy of Metal Finishing Association, Birmingham.)

6.5 Summary of the principal characteristics, advantages and disadvantages of surface treatment processes

Process	Characteristics	Advantages	Disadvantages
Surface cleaning			
Wire brushing, tumbling, barrel finishing, jet blasting	Mechanical methods of removing surface contamination, usually preparatory to surface finishing	Equipment simple, reliable and does not require skilled labour	Can be a dirty, noisy job Some parts too delicate to be cleaned by tumbling or wire brushing, and barrel finishing may require special holding fixtures Jet blasting not always able to reach internal recesses
Chemical cleaning, pickling	Organic and alkaline solvents used to dissolve and wash away soluble surface contaminants Pickling an acid etching process	Chemical solvents available to remove most greases, oils and other contaminants Awkward areas are easier to reach than with mechanical cleaning methods Process can often be totally enclosed Etching useful for both scale removal and controlled surface roughening before coating	Ferrous parts must be used as soon as cleaned to avoid rusting Thin sheet metal difficult to vapour degrease Special equipment and extraction facilities required to protect staff and environment Pickling only works if degreasing is carried out first
Ultrasonic cleaning	Controlled use of cavitation of the cleaning fluid	Cleaning characteristics can be easily changed and cavitation intensity readily controlled Delicate and intricate parts cleaned without problems Skilled labour is not required	Tank and transducer sizes control maximum part size that can be cleaned Some surfaces require drying after cleaning Gross contamination cannot usually be removed by this process
Surface smoothing			
Hand grinding, belt sanding, polishing/buffing, electropolishing	Refinement of surface finish by either abrasive or electrolytic means	Abrasive processes offer a wide range of surface finishes, even using unskilled labour. Most processes are portable Electropolished surfaces do not require further polishing or buffing and awkward shapes are not usually a problem	Most abrasive processes are labour intensive Electropolishing is only possible on electrically conductive parts and can be slow compared with abrasive polishing

Process	Characteristics	Advantages	Disadvantages
Surface protection			
Sacrificial protection	The provision of a metal that will corrode in preference to the surface to be protected	No significant surface corrosion until *all* sacrificial metal has been eroded away Sacrificial metal usually easy to replace Cheap process compared with alternatives	Only limited range of metals can be used – usually only zinc, cadmium or aluminium Many component shapes and applications not suitable for cathodic attachment or coating
Direct protection Conversion coatings (phosphating, chromating, anodizing)	Surfaces are chemically changed to provide the required corrosion-resisting characteristics	No increase in part size as surface is not coated but chemically changed Phosphating is an ideal preparation before painting steels. Also it enhances steel's deep drawing properties Anodizing ideal treatment for aluminium and its alloys and can be coloured if required	Phosphated surfaces have poor visual appearance, but this is not normally a problem as phosphating is usually used as a prepainting treatment Chromating is cheaper than anodizing but its wear resistance is inferior It also makes subsequent welding difficult All parts to be anodized must be submersible in an acid bath
Electroplating	Application of a totally enclosing metallic layer to the surface to be protected	Most metals can be plated or used as a plating material Plastics can be plated with suitable pre-treatment to make them electrically conductive Can be used as both a decorative and corrosion-resistant surface. Also used as a surface reclamation process for inspection gauges	Plating is a time-consuming process, requiring careful surface preparation, as all surface flaws persist through to the finished appearance Part must be electrically conductive (or made so), and suitable for immersion in a plating acid bath Plating does not produce a uniform coating thickness Some plating metals are prone to cracking and are porous, e.g. chromium
Painting	Application of an organic coating to the surface to be protected	Wide range of paints available, and most are easy to apply Good covering power. Many types of surface finish possible Can be incorporated into continuous metal strip production lines	Can be toxic so special operator protection usually required Some paints require special surface preparation and application techniques Most painted surfaces require at least two coats

168

Process	Characteristics	Advantages	Disadvantages
Vitreous coating	Application of a totally enclosing glass-like layer to the surface to be protected	Produces ceramic, porcelain or enamel coatings – ideal for high-temperature applications. Also offers good chemical and abrasion resistance Visually attractive finishes are produced	Prone to chipping Parts must be capable of being fired at high temperature without risk of damage or distortion
Vapour-deposited metal coating	Submicron coating applied by vaporising the coating metal and recondensing it onto the substrate	Thin coatings of most metals can be produced Ideal for precious metal deposition as so little is used Major uses are for coating cutting tools and in the food packaging industry	Capital cost of equipment necessary for this process is very high Parts must be enclosed in a vacuum chamber to be coated
Metal spraying (metallizing)	The spraying of atomized molten metal onto a surface	Ideal for reclamation of undersized parts Most metals and their alloys can be sprayed, so almost any desired surface characteristics are possible Any thickness can be built up Both portable and machine tool based units available	Surface tends to be slightly porous Process can be very noisy owing to supersonic nozzle velocities Surface preparation essential – usually by jet blasting or chemical etching
Cladding (diffusion bonding)	Covering sheet metal with other materials by surface atomic transfer (diffusion bonding) to form a protected metallic sandwich	Reduces anticorrosion treatment needed on assembled products Sheet materials can be made offering the precise surface characteristics required	Must be used the correct way round if coating is only on one side Edges tend to be potential corrosion sites Not a cheap raw material option, but costs may be more than offset by reduced number of surface treatment processes needed during manufacture

Bibliography

Anopol 1990. Stainless steel surface enhancement. *Surface Engineering* 6(3), 157–60.

Anopol 1992. Surface treatment enhances the life and appearance of stainless steel. *Stainless Steel Industry* **20**, 15–18.

Committee for the Promotion of Electroplating 1983. *A guide for designers and engineers*. Birmingham: Committee for the Promotion of Electroplating.

Freeman, D. B. 1986. *Phosphating and metal pre-treatment*. Cambridge: Woodhead-Faulkner.

Goodwin, T. & Fewtrell, G. 1992. *Coated steel: adding colour and durability. Chemistry now!* Cambridge: Hobsons Publishing.

Lambourne, R. 1987. Paint and surface coatings theory and practice. Chichester: Ellis Horwood.

Mattox, D. M. 1993. PVD: *vacuum evaporation – substrate fixturing, plating and surface finishing*. Albuquerque: Society of Vacuum Coaters.

Roessell, T. & J. Swain 1990. Effective chemical cleaning of stainless steel fabrications. *Stainless Steel Industry* **18**, 18–21.

Smith, W. F. 1990. *Principles of materials science and engineering*, international edition. New York: McGraw-Hill.

White, M. J. 1968. A survey of techniques for the vacuum deposition of thin metallic films. *Vacuum* **18**(12), 651–4.

Case study

A pump company produces a range of rotary positive displacement vane pumps (Fig. 6.25) that are used to pump lubricating oils of various viscosities. To extend the market for the pumps, pumping of low-viscosity water-based liquids was tested at pressures up to 25 bar. On the test bed the pumps were found to suffer a marked loss in volumetric efficiency (output/rev ÷ swept volume/rev), and this was frequently followed by their failing to pump at all after standing idle for more than a day.

What do you think caused this loss in pumping efficiency, and what suggestions would you make to solve the problem?

Solution

Diagnosis

From Figure 6.25 it can be seen that the pump's flat rectangular vanes slide in and out of their slots with each rotation of the vane rotor, spring pressure being applied at the root of each vane to keep its tip in permanent contact with the pump casing. When much thinner than normal liquids were pumped, the pressure that the vane springs applied was insufficient to provide an adequate fluid seal at the vane tips, and this caused the pump to deliver a reduced volume of liquid per revolution. Certain pumps ceased pumping altogether as a result of the vanes sticking in their slots. This was traced to surface corrosion on both vane surfaces and in rotor slots.

Recommended remedial action

While initial proposals might be to reduce the clearances between rotor slots and vanes, and to increase the tension of the vane springs, these suggestions do not address the underlying prob-

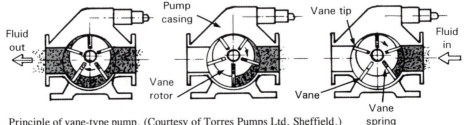

Figure 6.25 Principle of vane-type pump. (Courtesy of Torres Pumps Ltd, Sheffield.)

lem of internal corrosion. Furthermore, increasing vane radial pressure will increase vane/casing wear dramatically because the liquids being pumped no longer offer the internal lubrication provided by the lubricating oils normally handled by these pumps.

Clearly, surface protection of the pump internals is called for and, while changing the rotor to stainless steel will largely overcome problems with this component, for metallurgical reasons this is not an acceptable solution for the vanes. Because the smallest practical internal clearances must be maintained between rotor slots and vanes to minimize internal leakage within the pump, this makes it difficult to cover the vane's surfaces with a coating material; this would also increase all the vane's physical dimensions. Although such a general size increase could be allowed for before coating, achieving a perfectly uniform and controlled thickness over all vane surfaces would be extremely difficult in practice. There is also a risk that, in service, the coating could eventually become detached and either contaminate the liquid being pumped or damage the pump's internals.

Conversion coatings (§6.3.2.1) do not present such problems as they do not cause any increase in component size. Therefore phosphated bronze would be suitable for this application, and for low-pressure applications anodized aluminium may be an option. Vapour-deposited (§6.3.2.6) titanium nitride can also be used where appropriate, and while this is an actual coating the deposited thickness is usually less than $5\,\mu m$, and therefore has no significant effect on blade dimensions. Having made these changes to overcome the corrosion problems it may be no longer necessary to increase vane spring pressure, although if volumetric efficiency is still unacceptably low when pumping at higher pressures one may still be faced with a choice of increasing either spring pressure or pump speed.

Increasing rotor speed improves volumetric efficiency because leakage past vane tips is largely a function of the pressure difference that exists across vane tips and not their velocity. Therefore, increasing rotor speed increases the swept volume per unit time without a commensurate increase in leakage. It also increases the centrifugal force applied to the vanes, which will further assist fluid sealing at the vane tips although at the risk of increasing tip wear.

Questions

6.1 Retrieve the list of everyday items that have some form of surface protection that you were asked to produce in the exercise in §6.3. In the light of what you have learned by studying this chapter, re-examine each of the applications on your list and decide whether you still agree with your initial assessment of the method used.

6.2 What are the three principal reasons why component surfaces are treated?

6.3 Discuss the typical production engineering problems associated with the cleaning of component surfaces.

6.4 What is the difference between barrel finishing and tumbling, and why is the former as much a finishing process as a cleaning one?

6.5 Why should chemical cleaning never be embarked upon without considerable thought?

6.6 Vapour degreasing is not satisfactory for cleaning large, thin sheet metal parts. Why is this and what alternative methods would you recommend?

6.7 What is ultrasonic cleaning and from where is the cleaning force involved derived?

6.8 How does a protection coating differ from a sacrificial one? Give examples of where each would be most appropriate.

6.9 State the main reasons for finishing a product by plating. What factors determine successful plating?

6.10 What is the main disadvantage of chromium as a plating material for steel, and what can be done to overcome this problem?

6.11 What are the benefits of electrostatic painting compared with other painting procedures?

6.12 What is a paint, and how does it differ from a vitreous coating?

6.13 Painting can be dangerous, particularly spray painting. Why is this and what special precautions should be taken to minimize risks to both the environment and the personnel involved?

6.14 Suggest suitable protective coatings for the following applications and describe briefly how such coatings might best be applied:
- metallic milk containers
- marine fittings whose appearance is not of prime importance
- aluminium components that are to be dyed to improve their appearance
- the building up to size of a plain metal bearing that requires oil retention features.

6.15 Sputtering offers a number of advantages over other methods of vapour deposition coating. What are its main benefits and how are they achieved?

6.16 Component reclamation is just one use of the metallization process. What other equally important uses does it have? Why is it frequently an extremely noisy activity?

6.17 Diffusion bonding can be used to produce metal sandwiches tailor-made to suit the application. How is this possible? What precautions must be taken when using such clad materials to ensure that they function as the designer intended?

Chapter Seven
Powder metallurgy

After reading this chapter you should understand:
- (a) what the process of powder metallurgy is
- (b) where, when and why powder metallurgy is used
- (c) the principal features of metal powders and how they are made
- (d) what happens during sintering
- (e) what secondary processes may be applied to components after sintering
- (f) where powders consisting of individually porous grains are used
- (g) how and why metals of controlled porosity are produced
- (h) how conventional plastic injection moulding machines may be used for powder metallurgy component manufacture
- (i) how metal strip is made using powder metallurgy
- (j) the principal constraints when designing sintered components.

7.1 What is powder metallurgy?

In Chapter 2 various methods of producing engineering components from liquid metal were described. But there is another way that objects can be formed from raw material which, like molten metal, has no initial fixed shape, and that is by powder metallurgy (PM). However, unlike casting processes, liquid metal is not involved. The PM process can therefore be described as the art of making objects from metallic and non-metallic powders.

7.2 The basic principles of powder metallurgy

The PM process is simple to understand (Jones 1960), but not quite as easy to carry out because of the specialised equipment, knowledge and experience necessary to achieve consistent quality.

The process commences with the metallic and/or non-metallic powder(s) required to give the end product its desired properties being mixed together (blended) and then poured into a metal die. The powder mix is then compressed in the die under heavy pressure (compacted) to produce what is termed a preform. Finally, this preform is heated (sintered) for a finite period in a vacuum or controlled atmosphere oven. When it emerges from the oven the end product is a PM component (Fig. 7.1). Each of these essential operations is now examined in more detail.

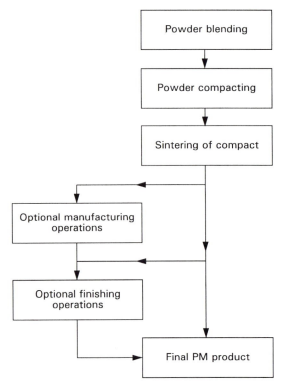

Figure 7.1 Basic steps in powder metallurgy.

Various optional post-sintering processes are also available to enhance the characteristics of the finished item if necessary (§7.2.5).

Note   A crucial point that should be clearly understood is that PM is *not* a casting (liquid metal) process, as the powder(s) do not melt during sintering (except in the special case of liquid phase sintering – §7.3.1).

7.2.1 Powder production methods and characteristics

There are numerous methods of powder production (Dowson 1994), involving chemical, electrolytic and mechanical disintegration processes. Another procedure is atomization, in which a stream of molten metal is forced through a small nozzle and then disintegrated by a jet of water or gas, depending upon the metal involved (Fig. 7.2). Fine-grained but irregular-shaped particles are produced by this method.

The most important parameters of any powder are its particle shape and size, particle size variation, purity and apparent density (mass per unit volume), all of which are significantly influenced by the way that they are produced.

7.2.1.1 Particle size and shape

The bigger the particle size, the more porous the end product. A fine grain size is therefore generally preferred as it also gives greater particle strength and so helps prevent particle fracture during compacting.

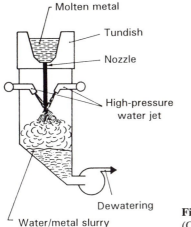

Molten metal

Tundish

Nozzle

High-pressure
water jet

Dewatering

Water/metal slurry

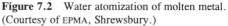

Figure 7.2 Water atomization of molten metal.
(Courtesy of EPMA, Shrewsbury.)

Shape is also an important factor as irregularly shaped grains compact better than truly spherical ones.

7.2.1.2 Particle size variation

If homogeneous (uniform) density and good compressibility are to be achieved in the compacting stage particle size variation must be kept to a minimum.

7.2.1.3 Purity

The higher the purity of the individual powders constituting a mix, the greater will be the metallurgical quality of the sintered products produced. However, high-purity powders are so expensive that only powders of the lowest acceptable purity should be used.

7.2.1.4 Apparent density (mass/unit volume)

Apparent density is a function of particle size and particle porosity, which together influence the degree of compressibility possible during compacting and hence the density of the final product. It is possible deliberately to make grains that have a porous structure, and these are used in infiltration liquid phase sintering (§7.3.1).

Common powders used in PM are pure iron (for special electrical and magnetic applications), iron + carbon + various alloying elements (to form different grades of steels, including stainless), aluminium, copper, brass and bronze (for filter elements and bearings). Exceptionally hard materials are made by mixing metal carbides (usually of tungsten, titanium or tantalum) with cobalt (British Standards Institution 1979).

7.2.2 Powder blending

Before the powder mix can be pressed into the required shape before sintering, it must be thoroughly blended to give a totally homogeneous mixture. A metal stearate- or graphite-based lubricant is usually added to reduce die wear and to minimize friction between grains during the compacting operation, so that the density of the preform produced is as uniform as possible.

7.2.3 Die filling and powder compacting

The powder mixture is poured into a hardened steel die of the requisite shape and is then subjected to one of two processes. It is either compressed under extremely high pressure, down to half of its original volume or even less, or it is vibrated to ensure complete die filling. The first of these options is called compacting and is carried out either cold or hot, while the latter option, for obvious reasons, is termed pressureless forming (§7.2.3.3).

7.2.3.1 Cold compacting

This is the most common process, and involves subjecting the powder mixture in the die to a high compressive pressure (between 150 and 1000 MN/m^2). Cold welding between the powder grains occurs, and gives the resulting preform sufficient green strength, as it is called, to permit its removal in one piece from the die. This avoids the need to put the die into the sintering oven, as is the case with pressureless forming (§7.2.3.3).

Compacting pressure can be varied at will, and has a major effect upon the tensile strength of the final PM product.

Double-acting presses are normally used as they provide a more uniform density preform, but even then a maximum length/width die ratio of 3:1 should not be exceeded if significant density variations are to be avoided (Fig. 7.3).

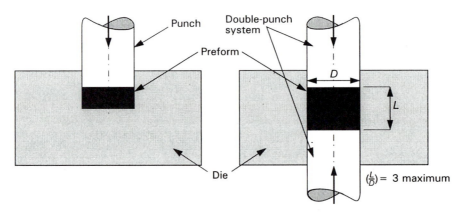

Figure 7.3 Single- and double-acting compacting punches.

7.2.3.2 Hot compacting

Some hard metals cannot be cold compacted satisfactorily and must be compressed hot. This has the disadvantage of having to be carried out in a protected atmosphere or vacuum to avoid oxidation problems. Not only must the die material therefore withstand typical sintering temperatures of 1000°C or more, but it also must be able to simultaneously cope with high compression pressures of approximately 800 MN/m^2. However it does have the benefit of combining both compacting and sintering into one operation.

7.2.3.3 Pressureless forming

Where porous parts such as filter elements are required, they are produced by loose sintering. The powder mix is poured into the die and is then vibrated to ensure complete die filling. Because it is not then compressed, as is the case with both cold and hot compacting, the absence of compacting pressure prevents the contents of the die from acquiring any green strength at all, and the powder must therefore be sintered while still in the die. As with hot compacting, this has the advantage of combining the compacting and sintering steps, but it again subjects the die to the rigours of high sintering temperatures.

Exercise Why do you think that the life of a die used in hot compacting is more arduous than one used in pressureless forming? Assume that sintering temperature is the same in both cases.

7.2.4 Sintering

The final essential step in the production of any PM component is sintering. This heating process coalesces (permanently fuses together) the powder mixture into a component of sufficient strength to carry out the job for which it was designed, and this is achieved by subjecting the preform to an elevated temperature for a significant time – up to 2 hours – usually in either a vacuum or controlled atmosphere.

Too high a sintering temperature will result in component distortion.

7.2.5 Optional post-sintering operations

As with normal wrought metals, properties of sintered parts can be modified or enhanced by the addition of supplementary processes. These can be in the form of either additional manufacturing operations and/or finishing operations.

7.2.5.1 Manufacturing operations

Following the sintering operation it is possible to machine PM components to achieve increased dimensional accuracy and to introduce features not possible during the compacting operation such as holes running normal to the direction of compacting.

The welding together of PM parts to produce a more complex-shaped end product is also possible (Fig. 7.4), fusion methods such as resistance welding (§5.2.2) being the most commonly employed procedures.

Figure 7.4 Complex multipart welded PM items. (Courtesy of EPMA, Shrewsbury.)

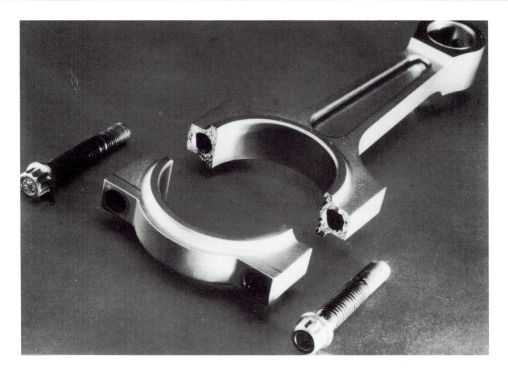

Figure 7.5 Powder forged BMW motorcycle engine connecting rod. (Courtesy of EPMA, Shrewsbury.)

It is also possible to forge sintered items (Fig. 7.5). This usually involves hot press forging (§3.3.3) of the sintered part in a closed steel or carbon die to achieve not only significant shape changes, but also a marked increase in material density. Increases in density and mechanical properties equal to, or even exceeding, that of traditional wrought metals are possible. This process is termed powder forging or sinter forging (Kuhn & Ferguson 1990).

7.2.5.2 Finishing operations

A wide range of post-sintering finishing operations are available and can be performed either immediately after sintering or following post-sintering mechanical operations. Typical processes that fall into this category are:

(a) *Coining*. The component is placed in a die (frequently the one used for the compacting operation) and recompressed. This results in a dimensionally more precise article, with accuracy levels of 4 μm/mm in the direction parallel to compacting and 2 μm/mm normal to the pressing line. This process is also known as sizing.

(b) *Impregnation*. Sintered material with 20–35 per cent controlled porosity, made by pressureless forming (§7.2.3.3), is usually employed in the manufacture of oil-impregnated (self-lubricating/oil-retaining) bearings. The component is first subjected to a vacuum to remove the air present within its microstructure. It is then submerged in oil and, finally, when the vacuum is destroyed, atmospheric pressure is restored, which forces the oil into the micropores of the material (usually bronze). Low-friction plastics are now also being used as impregnation materials because they offer a number of benefits, including improved machining characteristics and preventing the absorption of corrosive electrolyte during plating operations (see d).

(c) *Surface hardening and tempering*. Some ferrous PM components need to be hardened and tempered, and this can be achieved by gas carburizing followed by oil quenching. Salt baths, brine or water cannot be used because of their corrosive effects within the microstructure of the sintered material.

Exercise   Impregnation of sintered materials prevents unacceptable fluids from being absorbed into its microstructure. Why do you think it would be pointless to plastic impregnate sintered steel before surface hardening in the hope that water could be used instead of oil for quenching?

(d) *Surface coating*. Sintered materials can be surface coated satisfactorily by such processes as electroplating, galvanizing and chemical vapour deposition. As stated in (b), before electroplating can take place impregnation frequently has to be carried out to prevent electrolyte corrosion problems. Vapour deposition is used to apply wear-resistant surface layers of such materials as titanium carbide and aluminium oxide on modern carbide cutting tooling inserts.

Exercise   The stages through which a PM component passes during its manufacture are analogous to producing the culinary delight of a birthday cake! List the basic stages involved in the production of a PM component and suggest the equivalent steps in making your cake. (Do not forget the post-sintering options.)

7.3 Process variants

7.3.1 Liquid phase sintering

One way of achieving full (as wrought) material density in high-alloy steels ($7.85\,g/cm^3$), and hence the best possible mechanical properties, is to include within the powder mix one constituent that will melt during sintering. With cutting tool carbides this is usually cobalt, which melts at $1500°C$. Carbide mixes of hard metallic elements such as titanium and tungsten are sintered at temperatures slightly in excess of this, so during this operation the molten ingredient flows between the unmelted grains to act as a binder. (The melting point of titanium is $1700°C$ and that of tungsten $3400°C$.)

Coining is always carried out on liquid phase sintered products for the reasons stated in section 7.2.5.2(a). It also enhances the material's tensile strength and impact resistance. A variant of liquid phase sintering involves the use of powders in which the individual grains are porous. When the preform is sintered the melted ingredient still acts as the binder between the unmelted grains, but it also flows into the intergranular spaces by capillary action. This is termed capillary infiltration liquid phase sintering, and produces improved component density and strength.

Exercise   If you are not sure what capillary infiltration is place a small quantity of black coffee, not more than $5\,mm$ deep, in a saucer. Gently place a cube of sugar in the middle of this pool of coffee and watch what happens to the coffee. You have just witnessed capillary infiltration!

179

7.3.2 Metal injection moulding (MIM)

To mass produce small highly complex-shaped components, and to overcome the shape restrictions normally applicable to PM parts (§7.4), a process variant has been developed that replaces the normal die compacting step by injection moulding the preform (EPMA 1995). This is made possible by mixing very fine-grained metal powders (0.5–20 μm) with an organic (plastic) binder. The mixture can then be readily injected into a die of any required configuration using a conventional plastic injection moulding machine (§4.2.3). However, before the resulting preform can be sintered, it has to be de-binderized, leaving only the powder metal compact, i.e. the plastic binder is only introduced to provide the freedom to injection mould the powder mixture. Binder removal is accomplished by heating the preform in an oven. It is then transferred to a controlled atmosphere or vacuum furnace for sintering in the normal way.

Components made by the MIM method are rarely more that 100 mm long and have wall thicknesses between 0.5 and 5 mm.

7.3.3 Powder rolling

It is possible to produce rolled metal strip commercially by the PM. The metal powder, instead of being poured into a die, is continuously fed into a rolling mill, which then produces a preform in the form of an endless strip. The strip next passes through a sintering oven, and is then rerolled (a form of continuous coining), to produce the final strip material (Fig. 7.6). This process is particularly attractive for the production of small quantities of special metal mixes and for metals such as stainless steel that require a number of annealing steps if rolled from wrought material.

Figure 7.6 Sintered powder rolled metal strip. (Courtesy of EPMA, Shrewsbury.)

7.4 Design constraints

In spite of the wide design freedom offered by MIM, general PM component design has some major constraints that result from the way in which preforms are made. A basic limitation regarding preform length/width ratio of 3:1 has already been mentioned (§7.2.3.1), but there are a number of more general PM design restrictions that are now described and illustrated in Fig. 7.7a–f.

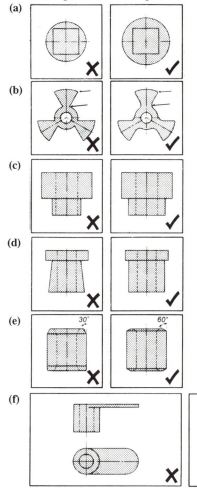

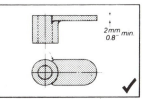

Figure 7.7 Main PM design constraints. (Courtesy of EPMA, Shrewsbury.)

Any portion of a component that is thin walled will probably need to be made in a thin-walled die, and this makes the die vulnerable to fracture. If high density and hence high compacting pressures are required, the risk of die damage is even greater.

As in good casting design (§2.7), sharp corners should be avoided. Sharp corners on compacts mean sharp corners in the die, and this again causes undue risk of premature die failure. Radii should therefore be used, although chamfers are preferred at component edges (see Fig. 7.7e).

Also in line with good casting design, PM components should not include large changes in section thickness. This can lead to significant density variations in the preform, and therefore weak areas in the sintered part. If major section changes are required in the finished part, then they should be machined to final form after sintering.

Re-entrant shapes cannot be made because the preform cannot be ejected from the die. If such shapes are essential the part must be made with parallel sides and then machined after sintering.

At component edges chamfers are preferable. However, to avoid chipping of punch edges they should be chamfered at approximately 60° (experience has shown that 45° or less is insufficient).

Components with a large surface area to thickness ratio are difficult to make because of density variations that occur during compacting. In addition, thin preforms are unacceptably fragile, and tend to crack during sintering. Design changes should therefore be made to eliminate such problems.

7.5 Advantages of PM

PM is the only way to produce certain material combinations. Indeed, a unique feature of this process is that, within limits, one can tailor the material composition to give the required prop-

181

erties to the sintered part. Not only is it possible to combine metals that will not freely alloy together, but metallic and non-metallic composite mixtures can also be produced. In addition, powder forging can be used to make metal parts that have densities as great as, or even greater than, their wrought equivalents. With both metallurgical content and density of the end product being so easy to manipulate, it is not surprising that this process is becoming increasingly attractive to today's design engineers.

Production rates can be very high, with powder filling and compacting cycle times as low as 1 second being common. This is achieved on automated flow lines, and carbide cutting tool inserts are typically made this way. Owing to this high level of mechanization unskilled labour is usually adequate.

There is normally no powder wastage during die filling and reject levels are usually low because most defects are detected at the preform stage before sintering (the powder can then be re-used). PM can therefore be an extremely cost-effective process: in some cases savings of up to 50 per cent are possible compared with alternative methods of manufacture.

Most sintered parts have a sufficiently good surface finish and accuracy that no post-sintering work is necessary. This is fortunate as some material combinations, such as diamond matrices and the hard metal carbides of tungsten and titanium, are exceptionally difficult to machine. Materials with a controlled level of uniform porosity can be made without difficulty and find particular application in the filter element and self-lubricated bearing fields. Metals used for filter element applications normally have porosity levels in excess of 27 per cent.

Small complex parts can now be made using conventional plastic injection moulding machines, providing batch sizes are sufficiently large.

7.6 Disadvantages of PM

Die costs can be expensive owing to the high-strength and wear-resistant steels necessary to achieve an economic die life. Therefore, depending upon component shape, metallurgical complexity and production volumes required, at least 15 000 parts are normally necessary before PM can be considered as a serious production alternative to casting, forging or machining from solid.

The maximum PM component size that can be made is limited by the press force necessary during compacting. Normally surface areas perpendicular to the axis of the applied load should not exceed $50 000 \, \text{mm}^2$.

Design limitations also influence whether PM is a suitable process, typical constraints being that no re-entrant shapes or cored holes normal to the axis of compaction are possible and component thickness/width ratio is limited to 3:1.

Sintered parts can be brittle, and generally the harder they are the more brittle they become, although coining after sintering helps to reduce this problem.

7.7 Some PM applications

Seventy per cent of all PM parts are manufactured for use in the automotive and oil industries (Fig. 7.8) (Dowson1994), typical applications being timing belt pulley wheels, alternator and distributor self-lubricating bearings, oil pump gears, starter motor and alternator brushes, gearbox selector forks, valve guides, wear-resistant camshaft cams and powder forged connecting rods (Fig. 7.5).

Figure 7.8 Examples of PM parts used in the oil industry. (Courtesy of EPMA, Shrewsbury.)

Submicron filter elements are made from porous PM metals, as are filter elements for filtering aviation fuel.

A major PM application is in the manufacture of tungsten carbide cutting tool inserts (Fig. 8.26b).

Metal injection moulding is used to produce large quantities of small items such as spectacle frames, watch cases and parts for textile machinery, computer peripherals and other office machines.

Composites such as those made from mixtures of metal and ceramic powders (cermets) are used in high-temperature applications typical of that found in aerospace products such as gas turbine engines and rocket motors.

Bibliography

British Standards Institution 1979. *Powder metallurgical materials and products, BS 5600*. London: British Standards Institution.

Dowson, A. G. 1994. *Introduction to powder metallurgy – the process and its products*. Shrewsbury, Shropshire: European Powder Metallurgy Association (EPMA).

EPMA (European Powder Metallurgy Association) 1995. *Guide to metal injection moulding*. Shrewsbury, Shropshire: European Powder Metallurgy Association.

Jones, W. D. 1960. *Fundamental principles of powder metallurgy*. London: Edward Arnold.

Kuhn, H. A. & B. L. Ferguson 1990. *Powder forging*. Princeton, NJ: Metal Powder Industries Federation.

Whittaker, D. 1991. Sintered materials – can they operate under fatigue loads? *Materials and Design* **12**(3), 167–169.

Case study

Although high-performance internal combustion engine connecting rods are produced from forgings, the con rods used in modern mass-produced engines are generally made from high-strength cast iron. Figure 7.5 illustrates a typical engine rod, although this one is made by powder forging (§7.2.5.1).

If you were a prototype design engineer and were asked to evaluate the possibility of mass producing sintered iron con rods, what important factors would you consider? Also, if it proved feasible, what advantages would such rods have over current manufacturing methods?

Solution

The first consideration would be whether the tensile strength of the required sintered material is adequate to cope with the anticipated peak applied loads. Furthermore, con rods have to withstand heavy cyclic loading in service, and as most mechanically stressed machine components that fail in service do so as a result of fatigue, fatigue strength is also very important and must be carefully examined.

While sintered materials have relatively low ductility, fatigue properties are generally found to be more consistently predictable than those of other engineering materials (Whittaker 1991). Therefore, providing the component is designed to be stressed below the material's fatigue limit, one can be confident that it will not suffer fatigue failure in service. If the fatigue strength of the sintered part is not adequate, it can be improved by hardening and tempering after sintering (§7.2.5.2c).

Note A material's fatigue limit is the stress level below which it will never suffer fatigue failure, irrespective of the number of stress oscillations imposed upon it. Note that some metals, such as aluminium, do not exhibit a definite fatigue limit.

The detailed shape of the con rod is extremely important. While the basic configuration is predetermined because it must perform in the same manner as existing rods and in the same working envelope within the engine, certain features such as section thickness require special attention to ensure uniform density and make PM dies as simple and robust as possible. Furthermore, fatigue stress is seriously affected by any design feature that causes a localized stress concentration area – sharp corners and holes (particularly sharp edged ones) are classic examples. It is therefore vital that all radii are as large and smooth as possible although this, of course, applies to *all* engineering components however they are made. Figure 7.5 illustrates such a con rod design. Other general design rules that must be complied with are illustrated in Figure 7.7a–f.

A benefit of using sintered con rods in preference to cast or forged ones is that mass will be somewhat less owing to the lower density of PM materials compared with equivalent solid metal, although this is partly offset by the slightly larger component required to withstand applied loads.

The high dimensional precision achievable with PM minimizes variability in con rod mass and balance. This also reduces considerably the machining requirements necessary compared with that required when using castings or forgings. This, in turn, reduces capital investment requirements when re-equipping for new engine manufacture.

Questions

7.1 What processes can be used to produce metal powders suitable for sintered products?

7.2 What is the effect of sintering on a powder metal compact?

7.3 Why is it necessary to compress powders in some PM applications and not in others?

7.4 Describe the basic operations involved in the production of a modern carbide throwaway cutting tool insert. What are the main physical properties that you would expect to find in such tooling?

7.5 Describe a suitable method of manufacture for a range of metallic oil filter elements.

7.6 A component is considered for manufacture by sintering. Discuss the factors that must be taken into consideration before deciding whether this process would be suitable. If the component could also be made by die casting, give reasons for choosing one process rather than the other. If the component is to be designed for manufacture by one of these two processes, outline some limitations of each.

7.7 State the advantages, disadvantages and limitations of sintering processes compared with those of closed-die forging.

7.8 Compare the scope of the powder metallurgy process with that of investment casting.

7.9 When would you consider using plastic injection moulding machines for the production of PM parts?

Chapter Eight

Shapes cut from the solid – material removal processes

After reading this chapter you should understand:
- (a) the processes of generating, forming and copying shapes using conventional machine tools
- (b) the essential features of the five groups of most commonly used basic machine tools
- (c) how material is cut using single- and multiedged cutters
- (d) how high-precision and ultrasmooth surfaces are produced using grinding, honing and lapping processes
- (e) why specific cutting pressure is important and how it is used to calculate cutting power requirements
- (f) the principal factors controlling the life of a cutting edge
- (g) the range of materials from which cutting tools are made
- (h) what is meant by hot hardness and its influence on tool life
- (i) why cutting fluids are used and their contribution to enhancing cutting efficiency
- (j) the operating principles and major benefits of non-traditional machining processes such as ultrasonic, water jet, electrochemical, chemical, electro-discharge, electron beam, laser beam and plasma arc cutting.
- (k) why machining plastics is so difficult.

In previous chapters of this book the production of components from molten metal (Ch. 2), by material deformation (Ch. 3), by fabrication (Ch. 5) and from powders (Ch. 7) have all been considered. However, there is another option available which may be used not only to produce components from raw material, but also to improve dimensional accuracy and/or the surface finish of items that have already been formed by one or more of the other processes mentioned above. This is called machining.

Machining is the controlled cutting away of unwanted material so that what is left is an item having the required geometry and surface finish. This many seem both simple and obvious, but it is one of the most complex and demanding areas of manufacturing and supports a multibillion pound industry.

8.1 Generating, forming and copying shapes

To cut material some form of relative motion is required between the cutting tool and the surface to be cut (the workpiece). This is the case irrespective of the type of cutting tool used or the shape required.

There is generally considered to be only three basic ways of producing a shape on a workpiece – by generation, forming or copying.

8.1.1 Shape generation

The most common and flexible method of creating a shape is by pure generation. The word pure is used to indicate that the profile produced results solely from the relative motion between cutter and workpiece, and is in no way determined by the shape of the cutting tool used.

The commonest forms of motion are circular and linear which, in isolation or in combination, are capable of generating a vast range of profiles. However, the bulk of generated shapes comprise either cylindrical or flat surfaces (Fig. 8.1a–c).

In the case of cylindrical machining (Fig. 8.1a), called turning, the new cylinder of reduced diameter is generated by traversing the cutting tool slowly along a line parallel to the axis of rotation of the workpiece.

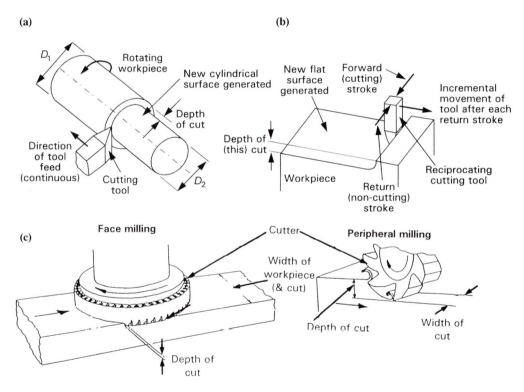

Figure 8.1   (a) Cylindrical surface generation (by turning on a lathe). (b) Flat surface generation (by shaping and planing). (c) Flat surface generation (by face and peripheral milling).

Exercise   What would be produced if either deliberately or accidentally a lathe cutting tool was traversed at an angle to the workpiece's axis of rotation? What type of surface do you think would be generated if the tool was traversed across the end of the job at right angles to its axis of rotation?

Exercise   What would be produced if a vee-shaped lathe tool's feed rate (axial movement /rev of workpiece) was much coarser than for plain turning? Assume that the depth of cut is not excessive? *Hint*: Remove the cap of a screw-top bottle and examine its neck!

The flat surface illustrated in Figure 8.1b is produced as follows. The cutting tool cuts in a straight line on its forward stroke and then returns to its starting position; before the next forward stroke commences, the tool is moved (fed) a small increment in the direction of the material still to be removed. This is called planing. If each incremental step is small enough a smooth surface will be generated; if each step is too great a flat but grooved surface will result.

Another method of creating a flat surface is by traversing a rotating (multitooth) cutter across the workpiece (Fig. 8.1c). Depending upon the form of cutter used, this is called either face or peripheral milling (§8.2.3).

It should be remembered that the geometry of the tools used in the examples of surface generation illustrated in Figure 8.1 have no influence on the shapes produced; *only the relative motion between tool and workpiece determines the profile.*

8.1.2 Forming

Unlike generation, in this case *the shape produced is determined entirely by the profile of the cutting tool*, which as a consequence is referred to as a form tool. Figure 8.2 illustrates how a form tool for turning is used to create a cylindrical shape that is a mirror image of its own profile. This is achieved by slowly feeding the form tool into the workpiece normal to its axis of rotation until the full form is produced.

A significant problem with forming is that if the perimeter of the cutting edge of the form tool is very long its feed rate must be slow. Even then there is still a risk of tool chatter (vibration), resulting in poor surface finish. Form tools are also used in slab milling, but unless the profile is simple the problem of chatter referred to above can be even more acute. Furthermore, form tools can be expensive to make and difficult to resharpen without damaging their profile accuracy.

8.1.3 Copying

Until the advent of computer-controlled machine tools (Ch. 9), profile copying was a popular method of repeatedly reproducing complex shapes. Its use is now declining, but for completeness the basic principles are briefly described here.

Copying involves tracing along the edge profile of either a sheet metal template (two-dimensional pattern) of the required shape or the surface of a carefully made prototype with a stylus. The resulting two-dimensional movement imparted to the stylus is then transferred either electronically or hydraulically to the cutting tool, causing it to mimic precisely the motion of the stylus.

When coupled with movement of the workpiece either rotationally (when turning circular shapes) or linearly (when copy planing longitudinal surfaces), the required three-dimensional profile is created (Fig. 8.3). As with pure generation, *the profile of the cutting tool has no influence on the shape produced.*

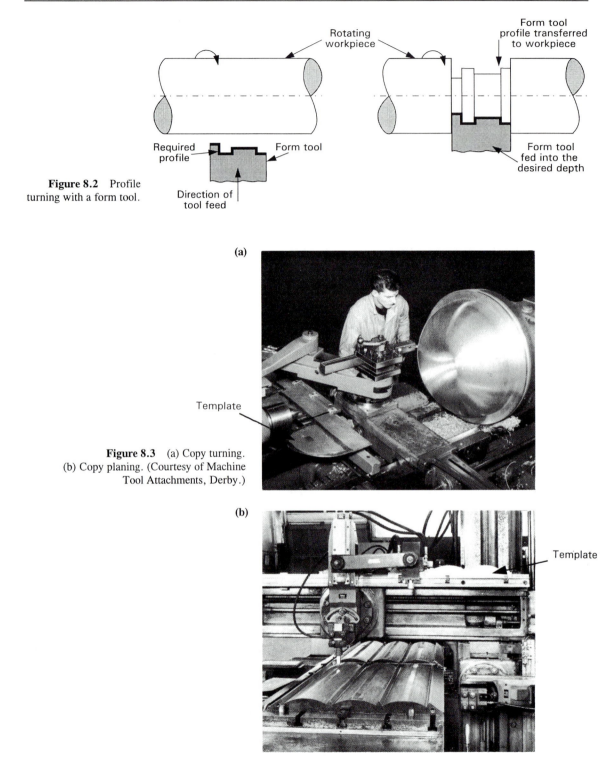

Figure 8.2 Profile turning with a form tool.

(a)

Template

Figure 8.3 (a) Copy turning. (b) Copy planing. (Courtesy of Machine Tool Attachments, Derby.)

(b)

Template

8.1.4 Production of more complex shapes

In describing the principles of profile generation, forming and copying, a range of shapes were considered, varying from simple flat and cylindrical ones to slightly more complex three-dimensional examples. However, within the limits of practicality, the complexity of shapes that can be made is limited only by the ingenuity of the production engineer.

Two shapes that fall into this more complex category and which are of vital importance in engineering are the involute and the spiral, the former being the basic profile found on gear teeth and the latter being the basis of all screw threads. While every method of gear and thread manufacture involves either copying, forming or generation in one way or another, their detailed description is beyond the scope of this book, but will be found described in both larger texts (Kalpakjian 1984, DeGarmo et al. 1988) and in works dealing specifically with these topics (e.g. David Brown Special Products undated).

Broaches are a complex type of form tool that are frequently used to transform circular holes into more complex shapes (Fig. 8.4).

(a)

(b)

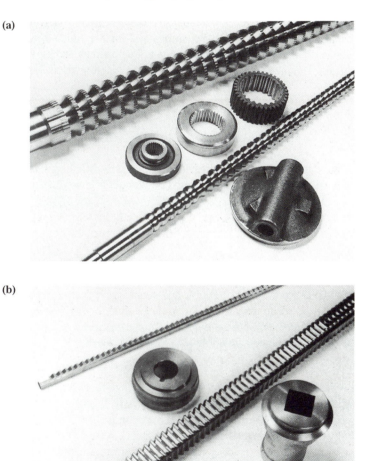

Figure 8.4 (a) Spline and round broaches. (b) Square and keyway broaches. (Courtesy of Redco Tooling Ltd, Redditch.)

191

In principle, a broach consists of a series of cutting edges whose first row of teeth fit into a premade circular hole, with the subsequent rows getting progressively larger and changing form over the broach's full length until the last row of teeth produces the final shape required. Thus, as the broach is pulled slowly through the original circular hole, each tooth removes a small amount of material such that, when the final row of teeth has passed through the hole in the component, the original hole will have been transformed progressively into the desired shape. Square, hexagonal, keyway and spline (precision parallel groove) shapes are the most commonly broached forms, although the fir-tree shaped roots of some turbine blades are also machined by surface, rather than circular, broaching.

8.2 Material removal by conventional machine tools

When discussing the production of workpiece profiles (§8.1) no description was given of precisely how the relative motion between cutting tool and workpiece is actually achieved in the workshop. Because precise control over the movement of both cutting tool and workpiece is essential to obtain an acceptable level of repeatable accuracy and surface finish of the end product, increasingly sophisticated precision machinery has been developed over the past 200 years to meet this demanding task. They are collectively referred to as machine tools.

Note Do not confuse machine tools with cutting tools! Cutting tools cut and machine tools control machining movements.

A wide range of machine tools are available, including ones controlled by computer (CNC machines, see Ch. 9), but only the most commonly used conventional machines are discussed here.

All machine tools have two tasks, and these are to hold both cutting tool and workpiece rigidly in the correct relative position to one another at all times, and to provide them with the necessary linear and/or rotary motion.

Note To specify in detail what machine tool(s), cutting tools and sequence of operations are required to make a given component in the most efficient manner is the job of the production (process) planning engineer. Process planning is a highly skilled task, requiring much practical experience and process knowledge, as well as detailed awareness of the machinery available (see also §9.2.1).

It is convenient to divide the diverse range of machine tools available into five basic families, the machines in each group being designed to perform broadly similar types of job. These five groups are lathes, shaping/planing, milling, drilling/boring and abrasive machinery. Only the principal features of each are now briefly described, as detailed descriptions of individual machine tools and their many variants in each group are not appropriate to the "basic principles" approach adopted in this book.

8.2.1 Lathe group

Two features are common to all machines in this family: it is the workpiece that is rotated and the general configuration is designed to cater for circular jobs where length is considerably larger than the diameter (i.e. a high L/D ratio). This is not to suggest that only parts with a large L/D ratio can be machined (turned), but long items such as shafts, special bolts, studs, etc. present no problem within the physical limits of the machine's working envelope.

The types of lathe available range from the basic but highly versatile centre lathe (Fig. 8.5) to more specialized variants such as capstan, turret and automatic lathes (Lindberg 1983), designed to increase productivity – usually at the expense of versatility. Highly skilled operators are required to realize the full capabilities of centre lathes, and are usually found in tool rooms (this is where tools, gauges and fixtures are made and maintained) and development workshops where one-off components are the norm.

The more specialized lathes are designed for high-quantity production: once set up by skilled staff they can then run unmanned or be operated by semiskilled workers.

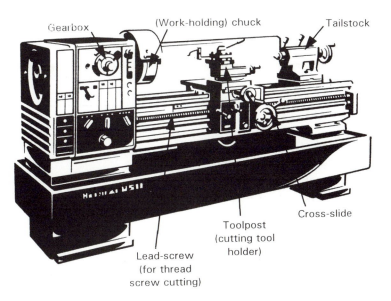

Figure 8.5 Elements of a basic centre lathe. (Courtesy of Buck & Hickman Ltd, Sheffield.)

8.2.2 Shaping/planing machines

A common feature of this family of machines is that either the cutting tool or the workpiece is reciprocated, a small incremental feed step occurring before each forward stroke (Fig. 8.1b). These machines are therefore specifically designed for generating flat surfaces.

With shaping machines the tool is the reciprocating member (Fig. 8.6a), but the problem with this arrangement is that there is a practical limit to the length of ram, and hence length of workpiece, that can be machined in this way. To overcome this limitation planing machines are used.

Planing involves precisely the same relative motion between cutting tool and workpiece as is used in shaping but, instead of the tool being reciprocated over the job, the job is reciprocated under the tool (Fig. 8.6b). This removes the restriction on workpiece length, as the floor-

(a)

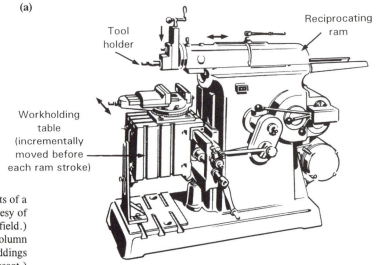

Figure 8.6 (a) Elements of a shaping machine. (Courtesy of Buck & Hickman, Sheffield.) (b) Large double-column planer. (Courtesy of Giddings & Lewis Ltd, Prescot.)

(b)

mounted reciprocating bed of the machine on which the component is clamped can be made as long as necessary. Beds 25 m long are available, whereas shaper ram stroke lengths are limited to no more than approximately 0.5 m.

Planer versatility can be enhanced by fitting a copying attachment similar to that described in section 8.1.3 and, while copying is generally in decline in the face of competition from multiaxis CNC machine tools, the cost of adapting an existing long-bed planing machine is only about 5 per cent of the cost of a new CNC machine of similar length. Indeed, certain parts of the roof of the Channel Tunnel were machined using long-bed planing machines fitted with such a copying device (A. Johnson, personal communication, 1994).

Planing productivity can be greatly increased by using a pair of cutting tools, one facing forwards and the other facing backwards. This enables cutting to be carried out on both the forward and return strokes, as opposed to the normal regime of cutting on the forward stroke only. The longer the machine tool's bed, the greater the benefits to be gained from such double-cutting motion.

8.2.3 Milling machines

Like shaping and planing machines, milling is also a machining process primarily suited to the production of flat surfaces rather than circular ones. However, shaping and planing machines (and lathes) all use single-point cutting tools (§8.3.1), whereas milling machines (and drills) use multitooth cutters (Fig. 8.7).

The essential relative motion between cutter and workpiece is achieved in this machine tool family by the cutter being rotated and the workpiece passing slowly under it (Fig. 8.1c).

The most commonly used types of milling machine are vertical and horizontal millers (Fig. 8.8a & b), vertical and horizontal referring to the plane of the cutter's axis of rotation. Vertical axis milling is usually called face milling and horizontal milling is termed peripheral or slab milling.

Figure 8.7 Single- and multiedged cutting tools. (Courtesy of Sandvik Coromant Ltd, Halesowen.)

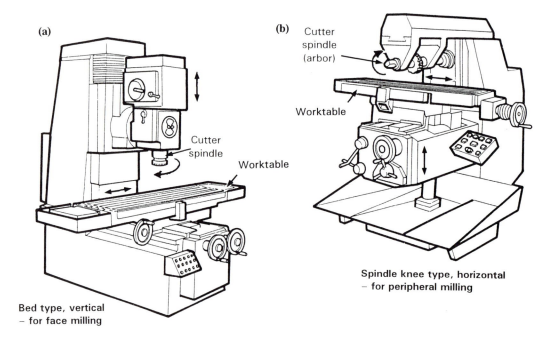

(a)

Cutter
spindle

Worktable

Bed type, vertical
– for face milling

(b) Cutter
spindle
(arbor)

Worktable

Spindle knee type, horizontal
– for peripheral milling

Figure 8.8 (a) Vertical miller. (b) Horizontal miller. (Courtesy of Sandvik Coromant Ltd, Halesowen.)

Note *Universal* milling machines are ones that are designed to allow for the cutting tool spindle axis to be oriented in either the horizontal or vertical mode as required. *Plano-millers* are machines that have the form of planing machines (long bed, etc.), but instead of using a single-point cutter one or more vertical milling heads are employed. This enables large stock removal rates to be achieved on components that are far too big to fit onto the bed of a normal milling machine.

Milling machines are the second most common conventional machine tool after lathes, and while used principally to generate flat surfaces contours can also be produced using suitably shaped form cutters (§8.1.2). Milled surface accuracy and finish are similar to those obtainable from lathes (50–75 μm R_A) and, providing large material removal is not necessary, a single cut can often produce the final shape to the required accuracy and surface finish.

8.2.4 Drilling/boring group

One of the commonest geometric features required in engineering is the hole. Drilling holes is probably the process best known to both engineers and non-engineers alike, as it is a regular activity in most DIY jobs. Although it is possible to drill holes using a lathe by rotating the workpiece and feeding a stationary drill from the tailstock (Fig. 8.5), all the machine tools in this family spin the cutting tool and feed it into a stationary workpiece.

The difference between drilling and boring is mainly a question of hole size. Although conventional drills of 50 mm diameter and more are available, above approximately 30–40 mm

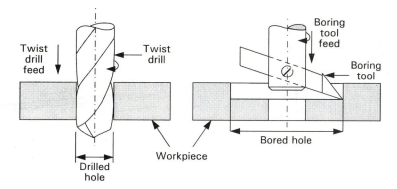

Figure 8.9   Difference between drilling and boring a hole.

they become more difficult to handle, so holes bigger than this are usually cut using a tool with a single cutting edge rotating at the radius of the required hole (Fig. 8.9). This is termed boring and is also the method used to produce larger internal holes and circumferential grooves in workpieces turned on lathes.

Drilling machines are available in a range of sizes, depending upon the maximum size of drill that they can hold, and range from small bench-type drills to large radial arm units capable of holding the largest twist drill sizes (Fig. 8.10a and b).

In addition to cutting deep holes, boring machines (Fig. 8.11a) are also capable of machining flat surfaces using either a single-point cutting tool held in a boring head and fed in a radial direction as it rotates (Fig. 8.11b) or a face cutter typical of that used on milling machines for face milling (Fig. 8.7).

The workpiece is clamped onto the worktable of the machine, and because the worktable can usually be swivelled accurately through any required angle it is easy to machine component faces and bore holes at precise angles relative to one another.

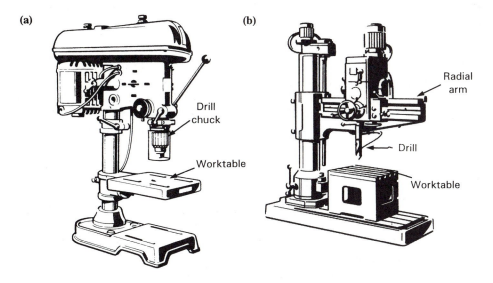

Figure 8.10   (a) Bench drill. (b) Radial arm drill. (Courtesy of Buck & Hickman, Sheffield.)

(a)

Figure 8.11 (a) Horizontal borer. (Courtesy of Giddings & Lewis, Prescot.) (b) Facing with a single-point tool. (Courtesy of Toolmex Polmach Ltd, Leicester.)

(b)

Tool

The most common angle is, of course, 90°, and is typically required when machining gearbox casings, in which input and output drive shafts must frequently be precisely at right angles to one another.

8.2.5 Abrasive machining group

All the machine tools in the previous four groups use single- or multiedge cutting tools to remove unwanted material. However, the machines constituting the abrasive family are rather different.

Most people are familiar with abrasives in sheet form (emery paper and sandpaper for instance), or as grinding wheels. While many abrasive operations are classed as surface smoothing processes (§6.2.1–6.2.3), there are a number of actual machining operations that employ this cutting medium, the principal ones being grinding, honing/superfinishing and lapping.

One feature of abrasive machining that sets it apart from all other forms of conventional cutting is that *it is the only one that can readily cut hardened materials*. Because so many engineering parts are hardened to extend their service life, abrasive machining is generally the only way to machine them to both the final precise size and surface finish required. It is therefore not surprising that heavy demands are usually placed upon most machine shops' abrasive cutting machines.

8.2.5.1 Grinding machines

Many parts require grinding as their final machining operation to achieve both high precision (5 µm) and finish, and because this usually involves removing a small volume of material from the surface of existing shapes it must be possible to grind virtually any profile produced by machine tools in any of the other machine tool groups. If this were not so there would be occasions when certain areas of a part could not be finish ground. As this is an unacceptable situation, the range of grinding machines that has been developed over the years is extremely varied. However, as with non-abrasive cutting, cylindrical and flat surfaces are by far the commonest profiles, and these are mainly catered for by external/internal grinders and surface grinders respectively.

External grinders are used for cylindrical shapes and are synonymous with lathes (Fig. 8.12a). Internal grinders are used for internal hole grinding and are synonymous with boring machines (Fig. 8.12b). Both external and internal grinders rotate the workpiece at modest speed and simultaneously traverse it slowly to and fro against a grinding wheel rotating at high speed.

Note *Universal* grinders are ones that can be reconfigured at will to perform either external or internal grinding. Can you see a certain similarity between this definition and the previously defined universal milling machine (§8.2.3)?

Surface grinders are akin to milling machines in that they generate flat surfaces. With the exception that the workpiece is not rotated, the surface-generating principle is the same as that described above for cylindrical grinding, i.e. the workpiece is traversed to and fro under a rapidly rotating grinding wheel (Fig. 8.12c).

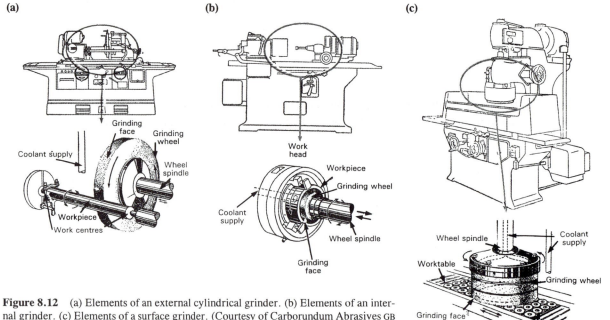

Figure 8.12 (a) Elements of an external cylindrical grinder. (b) Elements of an internal grinder. (c) Elements of a surface grinder. (Courtesy of Carborundum Abrasives GB Ltd, Manchester.)

Figure 8.13 Honing internal combustion engine cylinder bores. (Courtesy of Jones & Shipman plc, Leicester.)

There are a number of important specialist grinding machines, such as those used for grinding hardened threads and gear tooth profiles but descriptions of these are beyond the scope of this book.

8.2.5.2 Honing machines and superfinishing

Honing is usually a final machining operation preceded by grinding. It produces an extremely fine finish (1–3 µm) and can be applied to both flat and cylindrical surfaces, although its main use is in finishing high-precision circular holes such as internal combustion engine cylinder bores (Fig. 8.13) and the bores of hydraulic cylinders.

Honing machines range from manual units, where the workpiece is held in the hand and oscillated to and fro over a rotating honing head (the cutter) until the finish size and finish are achieved, to fully automatic machine tools. Automatic machines perform the same reciprocating motion, but usually it is the honing head that is traversed within the workpiece bore. They can be either horizontally or vertically configured, depending upon component length and whether multiple honing heads are used (Fig. 8.14a).

The simplest form of honing head used for bore finishing comprises a number of replaceable fine abrasive honing sticks that are either spring loaded or forced outwards on a taper to keep them in contact with the bore during cutting. Standard head sizes range from 6 to 400 mm diameter, although special ones are available up to 1 m diameter (Fig. 8.14b).

It should be remembered that, because the honing head is normally allowed to float in the bore, it will follow the existing line of this hole, and therefore cannot correct any hole positional errors. However, it will remove out-of-roundness, enhance surface finish and establish the bore's required finished diameter.

Superfinishing is fundamentally the same process as honing except that it is used to enhance finish on external rather than internal cylindrical surfaces, and on flat surfaces.

Like honing, this process involves rubbing an abrasive block over the surface of the workpiece, but in this case it is the component that is rotated.

The most commonly used configuration of this process is for enhancing external cylindrical surfaces. This involves pressing against the rotating workpiece a preformed abrasive block that is oscillating laterally at approximately 10 Hz (cycles/second) (Fig. 8.15). The whole cutting area is flooded with coolant, which also acts as a lubricant between abrasive and component surface. The finer the finish required, the finer the grade of abrasive block used. The whole process is so simple that it can be carried out on any centre lathe, the abrasive head unit being held in the machine's tool holder.

As with honing, superfinishing cannot correct eccentricity or out-of-squareness, but it will eliminate out-of-roundness.

8.2.5.3 Lapping machines

When a surface has been ground, and maybe even honed, its surface may still not be as smooth and scratch-free as required. If this is the case, the process of lapping is employed. Although lapping can be carried out on circular objects, it is normally associated with flat surfaces, the principle of the process being the same in both cases.

Very fine particles of abrasive powder (usually silicon carbide, aluminium oxide or diamond) are mixed with either thin mineral oil or water and spread over a soft metal surface called a lap plate (Fig. 8.16a). Cast iron is the most popular metal for this purpose. The workpiece is then placed on the lap plate and moved in a sweeping rotary motion, thus spreading the abrasive over the whole of the lap's surface. Because the lap plate must be softer than the workpiece material, the abrasive grains become embedded in the lap's surface and not the

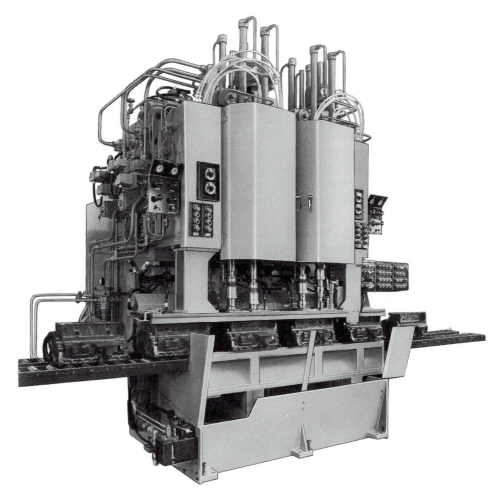

(a)

Figure 8.14 (a) Four-head cylinder block honer. (Courtesy of Jones & Shipman plc, Leicester.) (b) One metre diameter honing head. (Courtesy of Permat Machines Ltd, Coleshill.)

(b)

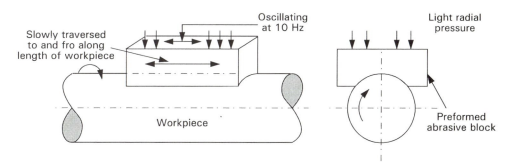

Figure 8.15 Superfinishing a cylindrical surface.

(a)

(b)

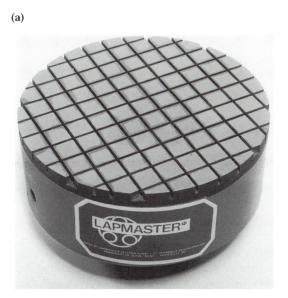

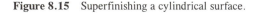

Figure 8.16 (a) Cast iron lap plate. (Courtesy of Lapmaster Ltd, Ivybridge.). (b) Pre-coated disk lap surface. (Courtesy of 3M United Kingdom Ltd, Bracknell.)

workpiece. Thus an abrasive layer is created on the lap, which both prevents lap wear and cuts the workpiece by removing its high spots.

Exercise What do you think would happen if you tried to lap a material that was softer than the lap – brass or aluminium for example?

It is particularly important that the abrasive powder used is extremely uniform. Otherwise, the larger grains would do all the cutting work and cause scratching of the component's surface.

Another way of creating an abrasive lap surface is to use a self-adhesive disc that is precoated with the required grade of abrasive material – usually diamond. The disc is stuck to a rotary base plate and then used for lapping as previously described (Fig. 8.16b).

Lapping is a slow process when carried out by hand, but if production quantities justify it lapping machines can greatly speed up the process.

Examples of applications for which very demanding levels of flatness and surface finish are essential are final finishing of the end faces of gauge blocks (§11.5.1) and the mirror-like faces required on metallurgical samples for macrographic examination (Fig. 8.16b).

Exercise   *It does not matter how much you read or how many line drawings and photographs you examine, there is no substitute for going into a machine shop and actually seeing at least some of the machine tools described in this chapter.* Try to make such a visit and, if possible, at least see a lathe, milling machine and grinder actually cutting metal. What follows will then be that much easier to understand.

8.3 Material removal processes

It is now necessary to examine precisely what happens when cutting tools used on conventional machine tools actually cut metal.

Most readers should, by now, have observed a lathe turning down (reducing) the diameter of a bar of material from D_1 to D_2 (Fig. 8.1a). Depending upon the material cut, the profile of the cutting tool and the cutting parameters used, the swarf (unwanted material) will have left the workpiece either as a mass of discrete chips or as one long chip.

While not so obvious, the cutting action of a grinding wheel is similar to a multitooth milling cutter in that the swarf is also in the form of chips (§8.3.2). These chips are, of course, extremely small compared with those produced during milling.

8.3.1 Single- and multipoint cutting

Most texts dealing with this topic refer to a cutting tool with one cutting edge as a single-point tool. This can be rather misleading as cutting tools have cutting edges not cutting points, and for this reason the more accurate terminology will be used from here on.

The cutting edge of a tool can be presented to the workpiece in one of two ways – orthogonally (two-dimensional) or obliquely (three-dimensional) (Fig. 8.17). For reasons that will become clear later in this section, most commercial cutting is oblique, but as the basic cutting principle is the same it is easier to examine in detail the less complex two-dimensional orthogonal process.

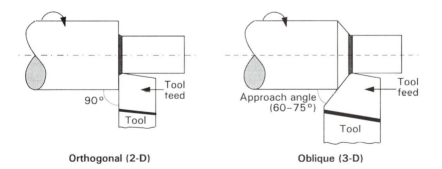

Figure 8.17   Orthogonal and oblique cutting.

When a tool's cutting edge is forced into contact with the workpiece material, the workpiece is subject to rapidly increasing shear stress. Eventually, the ultimate shear stress of the component material is exceeded and it fractures along the line of maximum stress.

Exercise   Can you see any similarities between the shearing action of a cutting tool and what happens when sheet metal is cut? If not, reread §3.7.1.1.

Obviously, the material from which the cutting tool is made (§8.7) must be much harder and tougher than the material to be cut; this prevents the workpiece cutting the tool!

The area of maximum stress (the fracture zone) is not a precise plane but is sufficiently close to it to be represented in this way, and is termed the shear plane (Fig. 8.18).

As the chip leaves the workpiece its underside rubs against the top face of the tool before either breaking into small chips or curling away in one long chip. This is an unfortunate phenomenon as it absorbs 30 per cent or more of the total cutting power for no gain to the cutting process, and tends to abrade the top face of the cutting tool. It should also be noted that the material removed from the workpiece becomes compressed during cutting, making the chip thickness (t_2) (Fig. 8.18) greater than the depth of cut (t_1). This absorbs yet more cutting power with no benefit to the basic process.

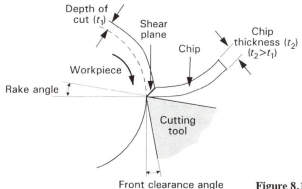

Figure 8.18   Cutting regime of a single-edged tool.

Thus, anything that can be done to minimize such unproductive areas of power absorption will significantly improve the overall cutting efficiency. As one of the most important factors influencing cutting power is the shape of the cutting tool itself, it is worthwhile examining its profile more closely.

If a piece of wood is pared longitudinally with a penknife, then, because the blade is so thin, the sliver removed (the chip) is not forced to curl upwards or rub along the sides of the blade. Thus, no significant friction occurs between the wood chip and the knife blade. Therefore, in the case of cutting metals with a normal single-edge cutting tool, the thinner the tool is the more efficiently it will cut.

Unfortunately, the thinner the cutting tool, the shorter its life before it becomes blunt or breaks. This is mainly because the cutting edge becomes excessively hot as the mass of metal in the tool is too small to conduct heat away from the cutting zone. It is therefore necessary to strike a practical balance between the thinnest possible cutting edge and one that will offer a reasonable working life. The profile of such a single-edge tool is shown in Figure 8.19. From this diagram it can be seen that (top) rake angle in particular has a major influence on the size of the tool's cross-section.

The majority of single-edge tools have a slightly more complex geometry than is illustrated in Figure 8.19. For example, two rake angles – side and top rake – are often ground on the upper face of the tool, and these combine geometrically to produce a compound angle called the

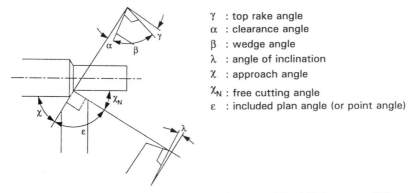

γ : top rake angle
α : clearance angle
β : wedge angle
λ : angle of inclination
χ : approach angle
χ_N : free cutting angle
ε : included plan angle (or point angle)

Figure 8.19 Principal angles of a single-edge cutting tool. (Courtesy of Sandvik Coromant, Halesowen.)

effective rake angle. In orthogonal cutting this is approximately the angle at which the chip leaves the top face of the tool.

With oblique cutting – the technique most often used commercially – it can be shown mathematically that an increase in the approach angle (Fig. 8.17) increases the effective rake angle (Boothroyd 1965). Because higher rake angles are preferable (see above), and since the cutting force required is reduced when spread over an increased cutting line length, oblique cutting is superior to orthogonal cutting. The direction of chip flow leaving the tool is also more desirable.

Exercise   Take a block of hard cheese, and with an ordinary kitchen knife (not a cheese knife), cut a slice with the knife blade *parallel* to the top surface of the cheese (similar to orthogonal cutting). Now cut off a slice with the knife blade *held at a 20–30° angle* to the surface of the cheese (similar to oblique cutting). You should notice a marked difference in the cutting force required.

Other factors influencing cutting power are cutting speed, tool feed rate and depth of cut. These are examined and quantified in §8.5.

Single-edge cutting tools are the basic tooling used in lathes, shaping, planing and boring machines, but for drilling and milling geometrically different cutters are used. The differences are not as great as one would at first imagine, because the basic cutting process previously described still applies. The major change is that these tools have more than one cutting edge, i.e. they are multiedge cutters.

The most familiar tool of this type is the conventional twist drill, in which two cutting edges are placed 180° apart on a common body (Fig. 8.7). Milling cutters generally have many more cutting edges (teeth) (Fig. 8.7), with at least two teeth cutting at the same time, as this helps to even out the cutting load placed upon the machine's drive system.

However, there is one cutting tool that has an enormous number of cutting edges: the grinding wheel.

8.3.2 Cutting using a grinding wheel

A grinding wheel may be considered as a cutter with many thousands of randomly positioned cutting teeth. Calling the cutting edges teeth is not quite accurate as they are actually randomly placed abrasive grains, but it does convey the principle of how metal cutting is achieved by grinding.

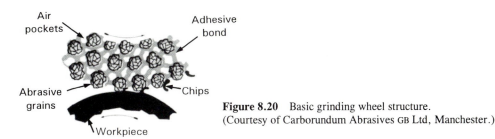

Adhesive bond

Abrasive grains

Chips

Workpiece

Figure 8.20   Basic grinding wheel structure.
(Courtesy of Carborundum Abrasives GB Ltd, Manchester.)

A grinding wheel consists of three components: abrasive grains, adhesive (bonding) material and air – the three As. How they are combined and how this influences wheel performance is discussed in section 8.7.2. Here only its cutting mechanism is described, and how this differs from single- and multiedge cutters.

The active (cutting) ingredient of any grinding wheel is its abrasive grains, and because they are uniformly distributed throughout the whole wheel there are always some suitably positioned on the wheel's surface to cut like small single-edge cutting tools. The fine crystalline nature of the grains used is such that the majority of grains will have at least one facet (face) optimally positioned for cutting (Fig. 8.20).

Material is removed in a series of chips (Fig. 8.20) and, because of their very small size, the cutting force makes them sufficiently hot that they normally glow a reddish yellow colour. These are the sparks that are often observed during grinding.

Like all cutting tools, individual active grains eventually become blunt, but it is unthinkable that the wheel should then be scrapped or that the blunt grains should be resharpened in some way. Therefore, the adhesive bonding material used is arranged such that when the cutting force necessary to make a blunt grain cut reaches a given level the grain will either fracture or pull free from the wheel's surface to expose sharp new cutting edges. This self-sharpening characteristic is referred to as attritious wear, its success depending upon the correct balance between bond and grain strengths in relation to the material being cut.

The coarser the grains used, the larger the chips produced, and hence the greater volume of material removed per unit time. However, coarse grains also produce rougher surface finish, indicating that a fine grit wheel should be used for finish grinding operations – often demanding accuracy levels of $\pm 5\,\mu m$.

Safety warning   *Grinding wheels are very dangerous* unless used with great care and adequate training. Indeed, it is a criminal offence in Britain (Factories Act 1961) even to fit a grinding wheel into a machine tool unless the person concerned has written certification that proper training in how to safely carry out such work has been given.

8.4 Specific cutting power and pressure

The three basic cutting forces involved in single-edge cutting are illustrated in Figure 8.21, and these are: tangential cutting force (F_c), axial feed force (F_f) and radial feed force (F_r) back along the tool shank.

The size and direction of the resultant cutting force (F) is a function of these three forces, although experimental tests have indicated that, in practice, the tangential cutting force (F_c) is far larger than either of the other two forces.

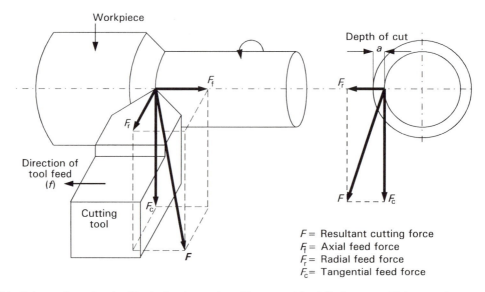

F = Resultant cutting force
F_f = Axial feed force
F_r = Radial feed force
F_c = Tangential feed force

Figure 8.21 Primary forces involved in single-edge cutting. (Courtesy of Sandvik Coromant, Halesowen.)

The power (P_c) required to actually shear a material is the product of tangential cutting force (F_c) and the linear cutting speed of the tool (V_c):

$$P_c = F_c V_c$$

where V_c is the linear cutting speed (m/min).

Note that for turning $V_c = \pi DN$, where D is the diameter of the workpiece (m) and N is the speed of the workpiece (rev/min).

If P_c is divided by the volume of material that can be removed per unit time (afV_c), the power required to shear unit volume of material per unit time (e.g. mm^3/s or cm^3/min) is obtained. This is termed the specific cutting power ($P_{specific}$):

$$P_{specific} = \frac{F_c V_c}{afV_c} = \frac{F_c}{af}$$

where a is the depth of cut (mm) and f is the feed rate (mm/rev or mm/stroke).

Because the units of $P_{specific}$ resolve to N/mm^2, this parameter is often called specific cutting pressure (symbol k_c). In reality, it is not a true pressure or stress figure, and it is claimed that the use of the word "specific" emphasizes this fact.

$$P_{specific} = k_c = \frac{F_c}{af}$$

Because k_c is approximately constant for any given material under set cutting conditions, it is a useful indication of a material's resistance to cutting, i.e. its machinability.

Note Traditionally in the cutting tool industry k_c is called a material's *specific cutting force* (Sandvik Cormorant 1994) (see Table 8.1), which is confusing to say the least, bearing in mind its units (N/mm^2)!

Table 8.1 Specific cutting "force" (k_c) for a range for metals (courtesy of Sandvik Coromant, Halesowen).

Material	Specific cutting "force" k_c (N/mm^2) *(not really a force)*
Plain low-carbon steel (C<0.25%)	2200
Plain medium-carbon steel (C<0.8%)	2600
Plain high-carbon steel (C<1.4%)	3000
Low-alloy steel, annealed	2500
High-alloy steel, annealed	3000
Stainless steel, annealed	2450
Cast steel, non-alloy	2200
Cast steel, low alloy	2500
Cast steel, high alloy	3000
Manganese steel, 12% Mn	4500
Rolled steel	4500
Titanium alloys	1530
Grey cast iron, low tensile	1300
SG iron	2100
Wrought and cold drawn aluminium alloys	800
Cast aluminium alloys	900

Table 8.1 lists values of k_c for many common metals, but the figures only apply to turning and drilling with coolant, a mean feed/rev of 0.4 mm and using a tool with an effective rake angle of 6°. Table 8.2 provides correction coefficients that cater for other feed rates in the range 0.1–1.4 mm/rev.

Because grinding is not generally a process used for large stock removal, but more a finish machining operation in which only small quantities of material are removed, specific cutting pressure (k_c) as a means of assessing material grindability is of little practical value. As stated in section 8.3.2, selecting the correct type of grinding wheel is far more important in determining how easily a material can be ground.

Table 8.2 Specific cutting "force" (k_c) correction coefficients.

Feed rate (*f*) (mm/rev.)	Correction coefficient
0.1	1.49
0.15	1.32
0.2	1.22
0.25	1.14
0.3	1.08
0.35	1.03
0.4	1.00
0.5	0.94
0.6	0.89
0.7	0.85
0.8	0.82
1.0	0.77
1.2	0.72
1.4	0.69

8.5 Machine tool power requirement

What is really required, of course, is the ability to predict the overall power requirement of a machine tool for any given cutting job. k_c can again be a helpful parameter.

The power (P_c) required to actually shear a material has already been shown (§8.4) to be given by:

$$P_c = F_c V_c$$

where V_c is the linear cutting speed (m/min). In addition, since $F_c = af k_c$, (§8.4), then

$$P_c = af k_c V_c$$

But P_c is not the total power required by the cutting process as it is based upon k_c, which, in turn, is based solely upon the tangential cutting force F_c. As this takes no account of such factors as power absorption in overcoming chip/tool friction, which can amount to 30 per cent or more of the total cutting power (R. Watson, Sandvik Coromant Ltd, Halesowen, personal communication, 1994), P_c will be a gross underestimate of the total power needed at the cutting tool.

To overcome this problem the cutting tool industry has therefore published k_c values that have been modified to take account of these other losses, i.e. using their k_c values gives power figures that are found to be correct in practice. The figures given in Table 8.1 are just such modified values and are typical of those used in industry.

Further correction factors are also required to take into account the overall mechanical and electrical efficiency of the machine tool (η), an average figure of 70 per cent being the industry norm.

Finally, to allow for cutting with semiblunt tooling, a further factor of 1.25 is usually included. Thus, the likely overall machine tool power requirement (P_{real}) in kW, for a given job, is:

$$P_{real} = \frac{\left(afV_ck_{c \text{ modified}}\right) \times 1.25}{0.7 \times 1000 \times 60}$$

$$\approx 30 \times 10^{-6}\left(afV_ck_{c \text{ modified}}\right) \text{ kW}$$

where a is the depth of cut (mm), f is the feed rate per rev or stroke (mm) and V_c is the linear cutting speed of cutting tool (m/min).

8.6 Tool life of single-edge tooling

If a cutting tool is used within its design limits, and does not fracture as a result of defects in the cutting tool material, its useful life can be estimated.

Note The term tool life does not necessarily mean the length of time from when it is first used until the tool is finally scrapped, but refers to the length of time that a cutting edge will continue to cut before it needs to be resharpened. However, as throwaway tips (8.7) are not resharpenable, their tool life does refer to the time until a particular edge is no longer usable. Do not confuse the length of time a tool is actually in the machine's tool holder with its tool life. This can be much longer than the time that the cutting tool spends actually cutting metal.

As with the previously described power calculations (§8.5), the tool life expectancy values calculated here are not exact, and only serve as an approximate guide. They are nevertheless still useful, particularly when predicting the likely effects on tool life of varying cutting speeds.

The algebraic relationship between tool life and cutting speed is known as Taylor's tool life equation, and is defined as:

$$V_cT^n = C \qquad \text{or} \qquad T = \left(\frac{C}{V_c}\right)^{\frac{1}{n}}$$

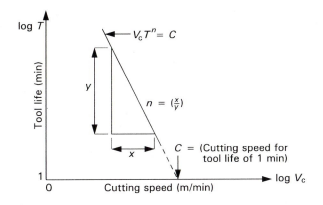

Figure 8.22 Taylor's tool life equation.

where V_c is the linear cutting speed of the tool (m/min), T is tool life (min) and n and C are constants.

Figure 8.22 is a logarithmic plot of the Taylor equation and illustrates precisely what these constants are and how they are calculated from plotted experimental results.

The index n is mainly a function of the cutting tool material used, and C is heavily influenced by the cutting tool material/workpiece material combination, as well as tool geometry and whether or not coolant is used. Their values are published by cutting tool manufacturers and, like the k_c values used in power calculations, are derived from carefully controlled experimental tests, with subsequent refinement to achieve improved correlation with normal production conditions.

The most important conclusion to be drawn from Taylor's relationship is that tool life is mainly a function of cutting speed rather than either depth of cut or feed rate. This is one of the most important laws of material cutting and clearly explains why, when removing large amounts of material (roughing out), feed rate and depth of cut can be increased without a major reduction in tool life. However, if cutting speed is significantly increased tool life shortens dramatically.

With finish turning, in which only small amounts of material are removed, high cutting speeds are normally used to achieve the required surface finish.

Note   It is untrue to suggest that feed rate and depth of cut have no effect at all on tool life and, while their effects are small in comparison with the effect of changing cutting speed, some cutting tool manufacturers quote a modified version of the basic Taylor equation that includes the effects of both the feed rate and depth of cut parameters.

8.7 Cutting tool materials

8.7.1 Single- and multiedge cutting tools

The properties of the material(s) from which cutting tools are made are crucial in determining how well the tools perform in use. The ideal cutting tool material would have the following properties and characteristics:

– high *hot hardness* (also called red hardness), i.e. it would retain its hardness when hot, as temperatures at the cutting edge can exceed 1200°C
– high toughness (high impact strength despite extreme hardness)
– high stiffness (high Young's modulus, E)
– complete resistance to wear and edge chipping
– good thermal shock resistance (will not crack if coolant is suddenly turned on in the middle of a cut)
– chemical inertness (no metallurgical affinity to workpiece material)
– low coefficient of friction (μ) between tool and workpiece materials
– high coefficient of thermal conductivity to conduct away heat from the cutting edge efficiently
– zero coefficient of thermal expansion (σ) so that the tool will not dimensionally change during cutting, making dimensional accuracy difficult to maintain
– easy to form into the required tool shape
– infinite life
– cheap.

Needless to say, no such material exists, and if it did it would certainly not be cheap! An increasingly wide range of tooling materials are, however, available, but, like most things in life, the more sophisticated the product the more it costs. The production engineer must therefore select the material that will best perform the required task at the lowest unit cost.

Probably the most important property of any cutting tool material is that it will retain its hardness at working temperature (its hot hardness). As this can exceed 1200°C in some cases (although 700–800°C is more usual), producing materials that can offer high hardness over such a wide temperature band is not easy (Fig. 8.23a). It is in this area that much cutting tool development work has been carried out.

Unfortunately, extreme hardness is usually accompanied by brittleness, so a compromise must be struck between adequate hot hardness and improved toughness, because the less brittle the tool the less likely it is to break prematurely in service (Fig. 8.23b).

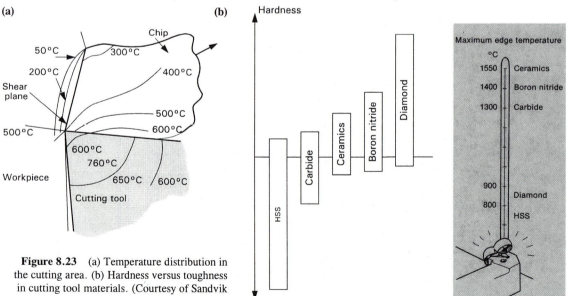

Figure 8.23 (a) Temperature distribution in the cutting area. (b) Hardness versus toughness in cutting tool materials. (Courtesy of Sandvik Coromant Ltd, Halesowen.)

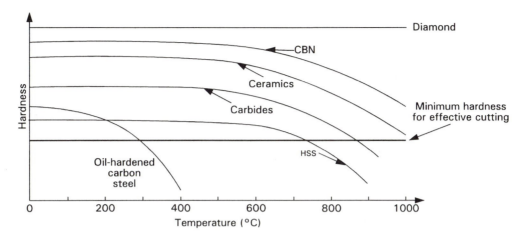

Figure 8.24 Hardness versus temperature for common cutting tool materials.

Figure 8.24 illustrates graphically the relative hot hardness of some of the most common tool materials in current use, although oil-hardened high-carbon steel is now hardly ever used because of its dramatic fall-off in hot hardness with temperature.

High-speed steel (HSS) is a steel that is, typically, alloyed with tungsten, molybdenum and either chromium or vanadium. Although it has been available for most of this century, it is still the most commonly used tool material, largely because of its ability to be formed into complex-shaped tools (Fig. 8.25) before they are hardened. Furthermore, it offers a good balance between hot hardness and toughness, and is readily sharpened many times during its useful life.

Figure 8.25 Complex-shaped tools made in HSS. (Courtesy of Sandvik Coromant Ltd, Halesowen.)

213

Exercise   Go into a DIY store and purchase a cheap set of small twist drills. The metal from which they are made will (probably) not be stated, but it will almost certainly be carbon steel, and the drills are likely to be black in colour. Also buy a set that are made from HSS – these will be shiny or golden in colour. Select the same size drill from each set and, with the aid of a hand-held power drill, make a hole in a piece of steel with each drill. Even using an oil lubricant, you should find that the carbon steel drill becomes blunt in seconds owing to its inferior hot hardness characteristics.

In Chapter 7 the process of powder metallurgy was studied, and it is to this technique that cutting tool manufacturers turned in the 1930s. The result was sintered tungsten carbide, which, as can be seen in Figure 8.24, offered a quantum leap in hot hardness compared with HSS. Indeed, the power limitations of the machine tools at that time became the controlling factor in the rate of material removal possible, rather than the limitations of the tooling material, as had been the case hitherto.

Unfortunately, there were two problems. First, even with the addition of cobalt as a binder (§7.3.1), the carbide was too brittle and fractured easily when subject to even modest impact cutting loads of the sort encountered when machining castings and forgings. Secondly, it was not possible to produce complex cutter shapes such as twist drills, broaches and form milling cutters.

As time progressed different powder formulations that significantly reduced the brittleness problem were developed, although increasing toughness did reduce hot hardness slightly. The severe limitations on shapes that could be made by powder metallurgy was not so easily resolved however, and even today, although the problem is less severe, similar restrictions still exist. For the bulk of carbide applications the cutting tool has a carbide insert or inserts, held mechanically or brazed onto a carbon steel shank or body (Fig. 8.7). A classic example of this approach can be seen by inspecting a typical masonry drill.

There is now an ISO system of grade classification of all carbide inserts (ISO 504:1975 Standards for Carbide Inserts), which is primarily based upon workpiece material suitability for a given carbide formulation rather than the characteristics of the carbide itself. There are three groups, designated P, M and K, which are further subdivided on a scale from 01 to 50 to indicate toughness and hardness; the higher the number the tougher the carbide, and the lower the number the harder it is.

In recent years tooling insert development has moved towards multilayer coated carbides offering enhanced toughness, shock resistance and longer life. The principal materials used for coating are titanium carbide, aluminium oxide and titanium nitride (Fig. 8.26a). Layer thicknesses are very small (typically $10 \mu m$) and are deposited by chemical vapour deposition (§6.3.2.6) methods. However, such coatings preclude resharpening of the cutting edge as the user cannot replace the coatings after sharpening, but even with uncoated inserts where this is not an issue it is still not economical to consider insert resharpening.

A wide range of insert shapes are available for use in turning (Fig. 8.26b), planing, milling, drilling and boring applications.

Many ingenious designs have been developed to allow inserts to be held rigidly in their shank/body but released readily. A typical example of a turning tool is shown in Figure 8.27. A major advantage with such tools is that an insert can be changed without the need to readjust tool height in the machine.

Ceramic inserts are made of sintered aluminium oxide and offer exceptional hot hardness and wear resistance. Unfortunately, they are more brittle than other sintered inserts, and their use is normally confined to high-speed cutting of cast and nodular irons, tough alloy and even semihard steels.

(a)

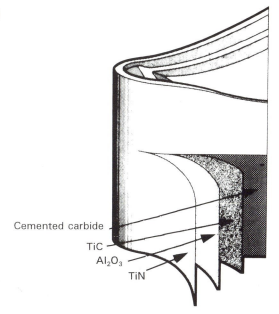

Cemented carbide
TiC
Al_2O_3
TiN

Figure 8.26 (a) Three coatings on a modern insert.
(b) Selection of turning inserts available. (Courtesy of Sandvik Coromant Ltd, Halesowen.)

(b)

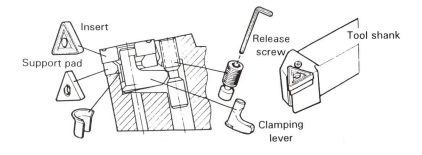

Insert

Support pad

Release screw

Tool shank

Clamping lever

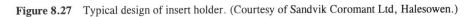

Figure 8.27 Typical design of insert holder. (Courtesy of Sandvik Coromant Ltd, Halesowen.)

215

Diamond is the hardest material known to man, so it should, in theory, make a good cutting tool material. Sadly, this is not the case, as it is extremely brittle and the slightest impact or variation in cutting force causes it to fracture. Its use is therefore severely restricted and is limited largely to high-speed, uninterrupted cutting of low shear strength materials such as aluminium, bronze and plastics (§8.10). Furthermore, it cannot be used to cut ferrous metals as there is an adverse chemical reaction between diamond and the iron in the metal.

Note To achieve the very high level of surface finish required on the aluminium discs used in PC hard disk drives, they are finish machined using a diamond tool.

Cubic boron nitride (CBN) was developed to overcome the problem of diamond being unsuitable for machining ferrous metals and, while not quite as hard as diamond, it is still harder than any other cutting tool material developed to date. As with all extremely hard materials, brittleness is still a problem, but as CBN is normally applied in the form of a tip that is brazed onto a carbide substrate (base) the cobalt in the latter gives a degree of toughness to the tip without adversely affecting its enhanced abrasion resistance at the temperatures encountered in fierce metal cutting situations.

8.7.2 Grinding wheels

Figure 8.20 (§8.3.2) illustrates that grinding wheels comprise three elements: abrasive, adhesive bond and air (the three As). Varying the type and quantity of abrasive and adhesive makes it possible to produce widely varying wheel cutting characteristics in a comprehensive range of sizes (Fig. 8.28).

The abrasives used are either aluminium oxide or silicon carbide and are available in a wide range of grit size from $250\,\mu$m to $2\,$mm maximum dimension. Silicon carbide is the harder of the two and, as such, is more brittle. It is therefore more suited to grinding low-strength materials such as cast iron, bronze, aluminium and non-metals. Steels are generally ground using aluminium oxide wheels.

The adhesive bond's task is to grip each individual grain until it has become so blunt as to be no longer effective as a cutting edge. They fall into one of two categories: vitrified (glass or similar material) or organic (resinoid, rubber or shellac). Because of the rigid nature of vitrified bonds, they are particularly suited to precision grinding, while the organic bonds are more flexible and therefore better suited to grinding operations in which slight flexing of the wheel might occur and where heavy-duty cutting typical of fettling castings is encountered.

When the job must be kept particularly cool, as in the case of grinding copper, or when an exceptionally fine finish is required, shellac-bonded wheels are preferred.

8.7.2.1 Wheel grade

Grade is the term used to specify the bond strength of the wheel, and is also often referred to as the wheel's hardness. The adhesive strength of the bond material is not the only factor that determines the degree of tenacity with which it grips the grains. The quantity of adhesive is also a major factor, as the more of each grain's surface area that is covered by bond, the stronger will be the grip that the bond will have.

The bond links between adjacent grains are called bond posts. The effect of increasing the amount of bond material on bond post size and strength can be seen in Figure 8.29a.

Figure 8.28   Typical range of grinding wheels. (Courtesy of Carborundum Abrasives GB Ltd, Manchester.)

A grinding wheel is said to be cutting soft when bond post strength is weak, and cutting hard when it is strong.

8.7.2.2 Wheel structure

Structure is the relationship between the number of grits and the amount of bond used and how close together they are packed. This is illustrated in Figure 8.29b. If the wheel is very densely packed adjacent grits will be too close together and the space between them will be too small to allow for each chip produced to be thrown clear of the wheel (Fig. 8.20). The spaces between adjacent grits will then become clogged, and the face of the wheel will no longer cut properly. The wheel is then said to be loaded, and the only way to remove this clogging is to redress the wheel.

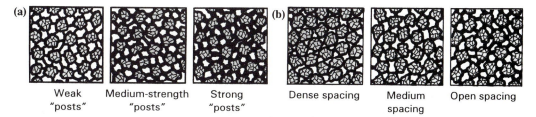

Figure 8.29   (a) Bond post strength. (b) Grit spacing. (Courtesy of Carborundum Abrasives GB Ltd, Manchester.)

Wheel dressing involves cutting away a thin layer of the working surface of the wheel by traversing an industrial diamond across it. This is clearly a waste of grinding wheel material, so the wheel should be changed for a more open-structured one to avoid clogging and to enable the correct self-sharpening action to take place (§8.3.2).

8.8 Cutting fluids

Taylor did not confine his metal cutting experiments to the study of tool life (§8.6), but he also discovered the benefits of flooding the cutting area with a coolant.

Modern machine tools are capable of delivering high levels of power to the cutting tool, much of which is ultimately dissipated in the form of heat through the chip, tool and workpiece. The adverse effects of high temperature at the cutting tool edge have already been discussed (§8.7.1), and heating of the workpiece makes accurate machining very difficult. Fortunately, 80 per cent of the heat generated during cutting is conveyed away by the chip (Lissaman & Martin 1982), but this still leaves sufficient heat to sustain high temperatures at the tool's cutting edge (Fig. 8.23a). Applying a flow of coolant at the tool/workpiece interface reduces not only tool/chip and tool/workpiece friction, but also the heat generated by the metal deformation that takes place during the shearing action of the cutting process. Surface finish is also significantly enhanced.

The thermal capacity (mass × specific heat at constant pressure) of water is more than double that of mineral oil, and this is one of the reasons that most coolants tend to be water-soluble oil compounds containing about 95 per cent water (Castrol 1991). Lubricating properties are adequate, and with the inclusion of extreme pressure additives one coolant is now suitable for most cutting duties.

In addition to carrying away heat from the cutting area, cutting fluids help flush away swarf from the immediate cutting zone. They also provide a thin protective oil film over both workpiece and machine tool surfaces.

With liquid coolants offering reduced cutting power requirements, cooler workpieces and extended tool life as a result of the lower operating temperatures, it is not surprising that most modern machining operations are carried out beneath a copious supply of cutting fluid delivered at pressures of up to 6 bar.

Safety warning   Generally coolant fluids are not injurious to health when used as intended, although prolonged exposure to the skin is neither recommended nor generally necessary. However, a small number of people are allergic to cutting oils, so if irritation or a skin rash is experienced medical advice should be sought (Castrol 1990).

8.9 Non-traditional machining processes

All the machining methods so far discussed in this chapter are called traditional machining processes, and they invariably involve, as part of their material removal process, the manufacture of arguably the most prolifically made but useless engineering product – swarf.

There are, however, a group of less familiar machining processes that employ totally different methods of material removal, and do not produce the conventional chip form of swarf. These chipless methods are usually called non-traditional machining processes (NTMs). Most of these processes have been devised to meet the increasingly difficult demands placed upon production engineers from both the technological and quality standpoints. Indeed, there are certain cutting procedures that were not even possible until some of these processes were developed, for example drilling a hole 150 times as deep as its diameter!

The range of NTMs available is continually expanding (Benedict 1987), but only the most important and popular applications are included here. They conveniently fall into four groups based upon the type of energy source they employ: mechanical, electrochemical, chemical and thermal.

8.9.1 Mechanical NTM processes

8.9.1.1 Ultrasonic machining

There are occasions when it is necessary to make holes or cavities in hard materials. A number of NTM processes offer this facility, but only ultrasonic machining can achieve it in electrically non-conductive materials such as ceramics and glass. Electrically conductive materials can also be ultrasonically machined of course.

This process relies upon a high-frequency impact action similar to a hammer drill. However, the tool does not actually touch the workpiece because it is not the cutting medium. Instead a flat-bottomed tool shaped as the mirror image of the required cavity or hole is flooded or submerged in a slurry of water and abrasive particles. The tool is then made to resonate at approximately 20 kHz in the direction of feed and, as it is lowered onto the workpiece, it hammers the abrasive grains against the surface to be cut (Fig. 8.30). This causes an eroding action sufficient to break or chip away the surface, leaving the required cavity. However, erosion can only occur if the material to be cut is hard enough to chip easily (above 60 Rockwell C).

The acceleration given to the individual abrasive particles is enormous and can result in an impact force on the work surface of as much as 150000 times the particle's weight (Benedict 1987). However, individual particle mass is so small that this still only results in a cutting force of 3–5 kg. Such low cutting forces make this process ideal for machining delicate components.

The tools used are normally made from ductile materials such as low-carbon steel, annealed stainless steel or brass: the harder the tool, the greater its wear rate.

Because this process cuts slightly oversize (equal to at least two grain sizes), tooling must be made undersize to counteract this. Surface finish is a function of abrasive grit size used and tool resonance amplitude – usually between 25 and 125 µm peak to peak.

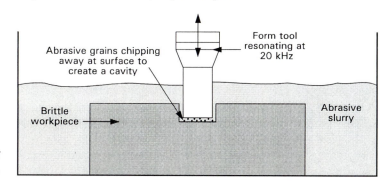

Figure 8.30 Basic principle of ultrasonic machining.

Holes from 75 μm to 50 mm diameter can be drilled with a length to diameter ratio of up to 40:1.

8.9.1.2 Water jet machining

Water may seem an unlikely cutting tool, but it has been in use as such since the early 1970s and, despite the high capital costs involved in water jet machining, the method is becoming increasingly popular, particularly in the automotive industry.

The process involves directing an extremely high-pressure jet of water – up to 4000 bar – at the material to be cut. With a suitable feed rate (providing the jet does not break up into a spray before it hits the workpiece, i.e. it sustains a coherent jet) the water will cut cleanly through a wide range of materials such as paper, wood, fibreglass and plastics (Table 8.3).

Table 8.3   Typical water jet cutting data. (Courtesy of Ingersoll-Rand Ltd, Manchester.)

Material	Thickness	Cutting speed (m/min)	Jet pressure (bar)
Water jet cutting data			
Fibreglass	4	2	3500
Vulcanized rubber	10	12	3500
Paper (75 g/m)	0.2	400	3300
Leather	5	20	3800
Plywood	5	10	3500
PVC	4	5	3500
Abrasive water jet cutting data			
Aluminium casting	6	0.46	2200
Low-carbon steel	3	0.43	3000
Stainless steel	14	0.15	3000
Titanium	6	0.38	3000
Glass	20	0.63	3000
Kevlar (carbon fibre)	13	0.38	3000

(a)

(b)

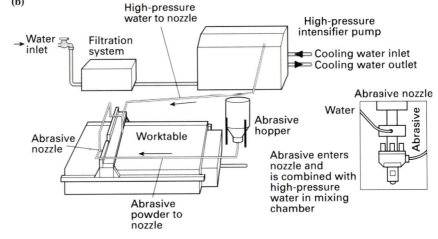

Figure 8.31   (a) 25 mm thick carbon steel cut by abrasive water jet. (b) Layout of typical abrasive water jet system. (Courtesy of ESAB Automation Ltd, Andover.)

As jet diameter is so small (70–500 μm) and water feed pressures are high, nozzle velocities of up to 800 m/s are reached (Mach 2.3).

When harder materials are to be cut (Fig. 8.31a), the cutting ability of the water jet can be greatly enhanced by introducing an abrasive powder, such as silicon carbide, into the jet stream (Table 8.3 and Fig. 8.31). This puts greater demands upon nozzle design if a reasonable life is to be achieved, and until recently the nozzles have been this process's Achilles heel. However, the latest designs incorporate exceptionally wear-resistant materials that have increased nozzle orifice life by as much as 2000 per cent (Benedict 1987) compared with traditional sapphire-lined ones.

Computer control of the nozzle head has also enhanced the capabilities of water jet machining, offering close nesting of parts for optimum material use (see Fig. 3.27) and making possible the cutting of complex three-dimensional parts.

8.9.2 Electrochemical NTM processes

Electrochemical machining (ECM) is based upon the same principle as electropolishing (§6.2.4), i.e. the reverse of electroplating.

The workpiece is wired as the anode (+) and the tool as the cathode (–), both being submerged in a conductive electrolyte. When a DC power supply is applied, metal is stripped from the workpiece by electrolytic erosion and moves towards the tool. However, before it reaches and then plates the tool, it is washed away in the electrolyte. To achieve this efficiently the electrolyte must be pumped between the tool and workpiece at high speed (15–60 m/s). Fortunately, as the gap between tool and workpiece is usually very small (25–750 μm), large electrolyte flow rates are not required, although pumping pressure is high.

To machine a component to the required profile using ECM requires a tool that has a mirror image profile of the desired shape. As the workpiece material is deplated the tool is progressively fed in the direction of erosion, keeping the gap between tool and workpiece constant, until the required depth is reached (Fig. 8.32a). Full 360° machining can also be achieved by making the tool in two parts and gradually moving them together, the workpiece material being sandwiched between the two tool halves. This is the way that certain turbine blades are finish machined (Fig. 8.32b).

As the tooling never actually touches the job, this process is ideal for machining fragile components. While only electrically conductive metals can be machined in this way, the process is also well suited to machining tough, difficult to cut metals with virtually no tool wear. This is illustrated in Figure 8.33, in which ECM is compared with a CNC machining centre (§9.2.3), which is frequently the only viable alternative. This graph also shows that ECM material removal rate is unaffected by the type of metal, its hardness or part complexity, and is largely a function of current density, which is typically of the order of 2–4 A/mm^2.

In addition to shape production, the ECM principle can also be applied to other conventional machining operations such as deburring, cutting and grinding (Benedict 1987). The application of ECM to these activities enhances both their productivity and their efficiency (McGeough 1974).

221

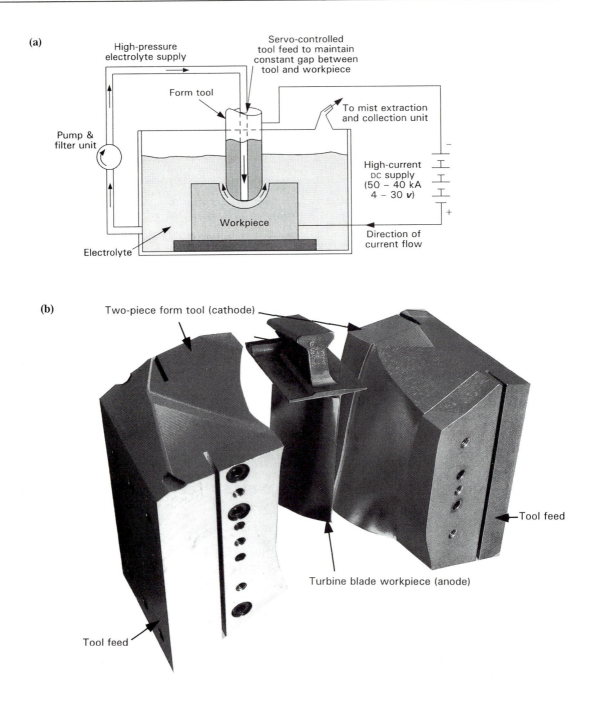

Figure 8.32 (a) ECM principle. (b) 360° machining of a turbine blade. (Courtesy of Amchem Co. Ltd, Birmingham.)

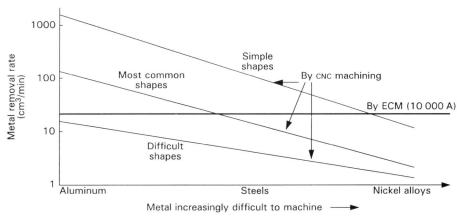

Figure 8.33 Comparison of metal removal rates (Anonymous 1985).

8.9.3 Chemical machining

Much of Chapter 6 deals with various ways of preventing surface corrosion and attack. A number of these procedures involve coating the surface to be protected with a corrosion-resistant coating. With the exception of sacrificial protection (§6.3.1), removal of a portion of this protective layer would result in corrosion of the exposed area. It is this situation that is exploited in chemical machining.

If a bare metal is either sprayed with or submerged in an appropriate acidic or alkaline solution, its surface will immediately start to dissolve in the chemical (chemical etching), and the longer the metal is exposed to the solution the more will be dissolved. However, if an area of the metal object is masked by a chemically resistant coating, only the unmasked metal will be etched away. This is the basis of chemical machining.

There are two main forms of chemical machining: chemical milling and chemical blanking. Both use the chemical etching process, but with chemical blanking etching occurs only along a precisely defined line until the etchant (the etching solution) eats right through the full thickness of the part.

Exercise   Compare this with mechanical blanking (§3.7.1.1), in which parts are made by mechanical cutting along a line as opposed to being chemically etched away along the cutting line. Can you think of an obvious advantage of chemical blanking delicate sheet metal parts?

With chemical milling the area to be etched is usually much wider than is the case with chemical blanking, and etching does not continue through the full thickness of the metal blank. It is therefore used for chemically machining recessed shapes of any form (Fig. 8.34), although the most common application is in the manufacture of printed circuit boards (PCBs). In this case the required copper tracks are masked, leaving the remaining copper coating to be etched away. Many PCBs are double sided, but this presents no problem for chemical etching as the process occurs simultaneously on all unmasked surfaces. This fact makes chemical machining a highly economical process in many situations, particularly when multiple parts are etched simultaneously.

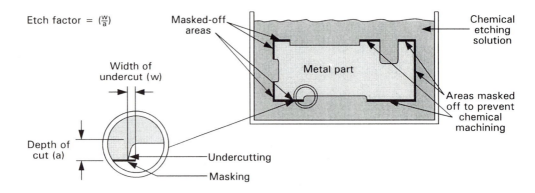

Figure 8.34   Chemical etching and the problem of undercutting.

One difficulty that does arise from this omnidirectional (in all directions at once) etching feature is the problem of undercutting. As Figure 8.34 illustrates, the etchant dissolves away the unmasked areas of metal, but it also, when the metal has been eroded below the mask thickness, eats away metal under the mask. To overcome this problem, masking must be modified so that the final shape required is what is actually produced. The ratio of undercut to depth of cut is termed the etch factor.

Exercise   How would you produce a tapered part?
Hint Time in the etchant controls the amount of metal dissolved.

8.9.4 Thermal NTM processes

In contrast to conventional metal cutting methods, thermal machining processes involve removing metal by controlled melting of the unwanted stock. This is why they are referred to as chipless machining processes. There are many such processes, although their only basic difference is the form and method of control of the thermal energy source used.

The four most common processes in this category are electro-discharge, electron beam, laser beam and plasma arc machining.

8.9.4.1 Electro-discharge machining – spark erosion

When electric arc welding was discussed (§5.2.1), it was shown that an electric arc is capable of generating sufficient heat to melt both electrode and workpiece. It is this principle that is also used in electro-discharge machining (EDM). However, there are some important differences between arc welding and EDM, one of the most significant being that with EDM the arc is intermittent owing to the pulsed DC power supply used. This results in a much less intense, and hence more controlled, arcing process.

In this case the objective is not to weld together the electrode and workpiece, but, in a controlled manner, to blast away material from the workpiece using thousands of discrete electrical discharges (sparks) per second (Fig. 8.35a) to create the required cavity. In practice, there is slight erosion of the tool (the electrode), but this effect can be minimized by adjustment of process parameters and selection of the most suitable tool material.

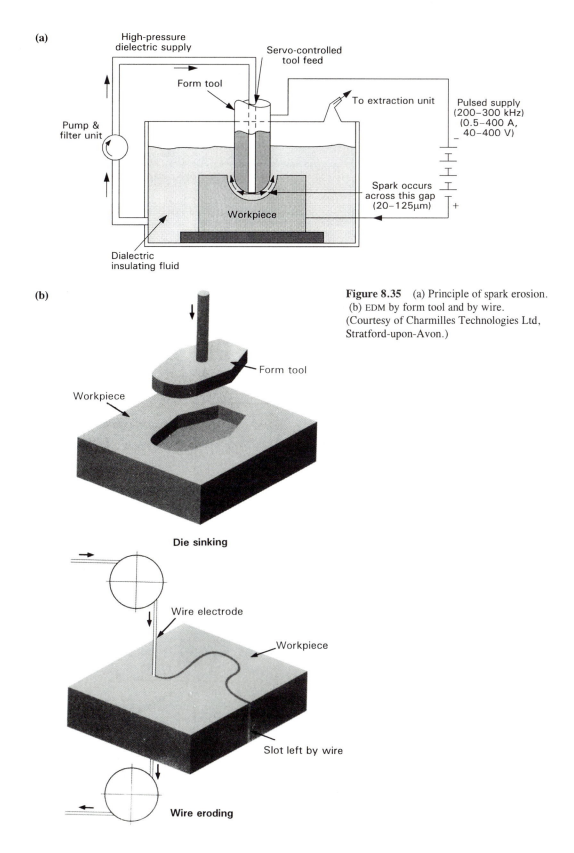

(a)

High-pressure dielectric supply

Form tool

Servo-controlled tool feed

To extraction unit

Pulsed supply (200–300 kHz) (0.5–400 A, 40–400 V)

Pump & filter unit

Spark occurs across this gap (20–125μm)

Workpiece

Dialectric insulating fluid

(b)

Form tool

Workpiece

Die sinking

Wire electrode

Workpiece

Slot left by wire

Wire eroding

Figure 8.35 (a) Principle of spark erosion. (b) EDM by form tool and by wire. (Courtesy of Charmilles Technologies Ltd, Stratford-upon-Avon.)

The gap between the tool and workpiece is flooded with an insulating fluid called a dielectric, which both aids the erosion process and flushes away the metal particles removed from the workpiece as the tool is progressively fed into the surface of the job.

By adjusting the energy and frequency of the electrical discharge and the spark gap between tool and part, the volume of metal removed and surface finish obtained can be closely controlled.

A wide range of shapes can be cut into the workpiece by this version of EDM because all that is required is a tool that is a mirror image of the required profile (similar to ECM tooling, §8.9.2). The most common tool materials are copper, brass and graphite, but if production quantities justify the cost copper tungsten is sometimes used as its erosion rate is very low. Unfortunately, it is nothing like as easy to machine as the softer metals.

As with ECM, because the tool never touches the job, easy to machine materials such as copper and graphite can be used to cut metals of infinitely greater hardness, and without imposing any significant cutting forces. This accounts for the great popularity of EDM, particularly in the field of die making (die sinking; see §2.8, die casting, disadvantages). Also, like ECM, only materials that are electrically conductive can be spark eroded.

An increasingly popular form of the EDM process is the replacement of the shaped tool with a thin circular wire 50–300 μm thick, whose electrode motion is akin to a band-saw blade (Fig. 8.35b). The major difference is that the electrode wire is not an endless loop, as it is in a band-saw, and the wire is scrapped once it has passed through the spark region. This is because, having been subject to the intense and hostile conditions within the erosion area, the wire is no longer circular or of a precise known diameter. Figure 8.35b illustrates the difference between this EDM variant and die sinking using a shaped form tool.

Most wire eroders as they are called are fitted with a computer-controlled worktable, which enables precise manipulation of the workpiece as the wire electrode passes through the slot. The spark area still needs to be well flooded with dielectric, a suitable stream normally being delivered from a supply tank via a nozzle clamped adjacent to the erosion area – where the wire passes through the slot in the workpiece.

Safety warning   The dielectric fluids used in EDM tend to give off hydrocarbon vapour, which is both inflammable and poisonous. It is therefore essential that such fumes are removed by a suitable extraction system.

As with EDM die sinking, the wire used in EDM wire eroding also does not touch the job, a radial job/wire gap being usually maintained at 25–50 μm. Cutting speeds vary with material thickness and not with complexity of the shape cut, but they are not high. A typical speed for 25 mm thick steel is 1–2 mm/min, depending upon the surface finish required, but this falls to just 0.3 mm/min when cutting 75 mm thick steel.

8.9.4.2 Electron beam machining

In Chapter 5 the use of a high-power electron beam to join metals by fusion welding was described (§5.2.1.5). The same energy source can also be used as a cutting tool, drilling being the form of cutting most suited to this process.

While the source of thermal energy is similar to that used in electron beam welding, there is a fundamental difference: with electron beam machining (EBM) the electron beam is pulsed rather than steady. Pulse frequency is adjustable from single-pulse operation up to a pulse rate of 1000 pulses/s. As with EBM welding, when using this energy source, the workpiece must be

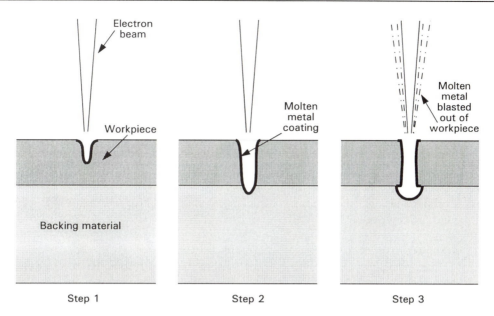

Figure 8.36 Principle of electron beam drilling. (Courtesy of Messer Griesheim Ltd, Cramlington.)

enclosed in a vacuum to prevent electron beam scatter.

EBM power intensity levels are little short of staggering, exceeding $1\,\text{MW/mm}^2$ (equivalent to the power of 500 electric kettles directed at just $1\,\text{mm}^2$ of workpiece area). With such energy levels any material can be instantly melted whatever its melting point.

To achieve a cleanly drilled hole in metal, experience has shown that a suitable non-metallic backing material is required for the following reason (Messer Griesheim undated). When the workpiece surface is hit by the electron beam (at half the speed of light), it is instantly melted and vaporized. This forms a vapour-filled hole surrounded by a thin layer of molten metal (Fig. 8.36, step 1). When the beam has penetrated the full thickness of the job it then hits the backing material and starts to also vaporize this as well (Fig. 8.36, step 2). Because the backing mate-

Figure 8.37 11766 holes 0.81 mm diameter drilled in 40 minutes. (Courtesy of Messer Griesheim Ltd, Cramlington.)

rial is carefully chosen to have a high vapour pressure, it escapes up through the hole in the metal, and in so doing carries with it the molten metal that is lining the hole in the workpiece (Fig. 8.36, step 3). This leaves a clean burr-free hole of the required size.

One pulse is usually sufficient to drill a hole, so the rate of hole drilling in a component is controlled by the speed with which the beam can be repositioned for its next hole (usually by computer control), rather than by the beam's drilling speed. Indeed, an extremely large number of holes per minute can be drilled using EBM. A good example of this is the cobalt alloy spinning head used in the manufacture of fibreglass and shown in Figure 8.37. Here 11766 holes 0.81 mm diameter are drilled in just 40 minutes.

EBM can be used to drill any material from metals to ceramics and plastics. Hardness is of no significance, and because drilling is so quick and localized no thermal distortion of the workpiece occurs. Material thickness is limited to approximately 10 mm and length to diameter ratio can be up to 15:1.

8.9.4.3 Laser beam cutting

Unlike EBM, because laser beams have the characteristic of being virtually parallel, with little tendency to scatter (diverge) in normal atmosphere, working in a vacuum is not necessary.

Both CO_2 gaseous and Nd:YAG solid lasers are used, the former being the most popular for general cutting and drilling, while the latter are used mainly for cutting out complex shapes and for high-precision drilling.

The parallel laser beam can be readily focused down to a sufficiently small focal point diameter (100–300 µm), to generate energy intensity levels not far short of those possible with an electron beam, and virtually any material can therefore be laser cut.

A laser beam energy source is extremely flexible in its range of potential applications, and in addition to welding (§5.2.1.6) it is extensively used for small hole drilling, two-dimensional and three-dimensional profile cutting, surface hardening and cladding. Here only the first two of these applications are considered.

(a)

(b)

Figure 8.38   (a) Two-dimensional profile laser beam cutting. (b) Oxy-fuel plate cutting (four heads.) (Courtesy of Messer Griesheim Ltd, Cramlington.)

Laser beams remove material in a broadly similar manner to an electron beam except that no backing material is required. Holes are usually drilled with a single pulse from the laser, the mixture of vaporized and molten material being expelled in a fine droplet spray. Holes so drilled are not geometrically precise when a CO_2 laser is used, although much better results are possible with the lower powered Nd:YAG lasers (up to 1 kW).

Holes with a length to diameter ratio of up to 100:1 are possible, depending upon hole size, although holes above 1 mm diameter are difficult because the larger the diameter of the beam's focal point the weaker is the intensity of the beam.

Profile cutting (Fig. 8.38a), in addition to using the laser beam as the cutting tool, also uses a high-velocity gas jet to assist in removing the gaseous and molten "swarf' and to speed up the rate of cutting. Air, oxygen and argon are the most commonly used gases. Profiling is invariably achieved by controlling the relative motion between the laser beam nozzle and the workpiece by computer.

The kerf (width of the cut) equals the beam's diameter at the focal plane, so little material is lost. Also, the beam can enter at any angle and can cut in any direction.

The type of power laser used, the assisting gas, the material to be cut and its thickness all influence the rate of cutting that is possible. For example, when cutting 2 mm thick low-carbon steel with a 1 kW CO_2 laser and oxygen as the assisting gas, the cutting speed is about 8 m/min, but with a thickness of 10 mm this drops to 1 m/min. This is why traditional oxy-fuel gas flame (Fig. 8.38b) or plasma arc cutting (§8.9.4.4) is a better option when cutting thicker metal plate.

Safety warning   Both electron beam and laser beam energy sources generate dangerous levels of electromagnetic radiation. While all commercial machines are well shielded to protect the operator, care must always be exercised by personnel using such equipment to ensure that safety procedures specified by the equipment manufacturers are adhered to at all times.

8.9.4.4 Plasma arc cutting

This thermal cutting process uses as its energy source a stream of extremely high-temperature gas. This is generated by sustaining an intense electric arc between a non-consumable electrode and the workpiece in a flow of gas, similar to plasma arc welding (§5.2.1.4). This puts an immediate limitation on the process in that only electrically conducting materials can be plasma cut; however, as virtually all metals are conductive, this is not usually a serious problem.

To generate and sustain the plasma arc, a pilot arc (Goodwin undated) is first struck between electrode and nozzle (*not* the workpiece), to give a jet of conductive ionized gas (Fig. 8.39a). When the nozzle is brought near enough to the workpiece the arc automatically transfers from

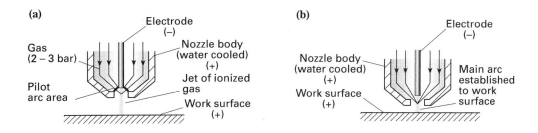

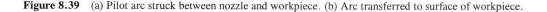

Figure 8.39   (a) Pilot arc struck between nozzle and workpiece. (b) Arc transferred to surface of workpiece.

the nozzle to the workpiece (Fig. 8.39b). Power is then increased and the main arc is fully established.

The energy intensity of a plasma gas stream is sufficient to melt the workpiece rapidly at the desired point, molten metal being removed from the cutting zone by the high-energy ionized gas stream (Goodwin undated).

Various combinations of electrode and gas are used, but conventional low-cost systems use air at 2–3 bar as the gas, and either a zirconium or hafnium water-cooled electrode. Electrode life is a function of the number of cutting starts, 250 being a typical life figure. This usually means between 2 and 10 hours of actual cutting life, depending upon the length of cutting between starts.

Figure 8.40 illustrates typical low-carbon steel cutting rates using a 30 kW air plasma unit as a function of plate thickness, and a comparison is also shown with cutting speeds possible using conventional oxy-fuel cutting (Goodwin undated).

Materials up to 150 mm thick can be cut using a plasma arc, but the bulk of applications lie more in the 5–75 mm range. Kerf width is much greater than with laser cutting and is up to double that produced when oxy-fuel cutting. Widths of 2.5–6 mm are typical.

Sharp corners are difficult to produce, and a significant HAZ (§5.2.1) occurs at the cut edge, although little deburring is necessary. Owing to the conical form of the plasma jet, cut edges tend to be slightly bevelled (Fig. 8.41), although not to an unacceptable degree in most cases.

As with EBM and laser cutting, most modern plasma arc cutting installations employ computer control for precise manipulation of movement between torch and workpiece.

An example of plasma arc applied to tube cutting is shown in Figure 8.42.

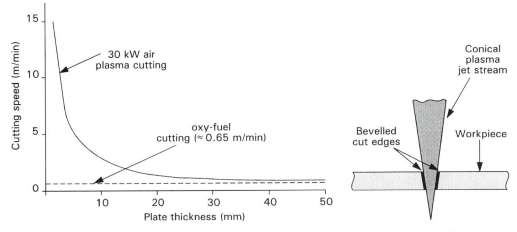

Figure 8.40 Cutting speeds of plasma arc and conventional oxy-fuel cutting.

Figure 8.41 Bevelling effect of a plasma arc.

8.10 Machining plastics

Because of certain inherent properties of plastics they are all difficult to machine.

When any material is machined by conventional cutting methods, heat is generated at the tool/workpiece interface. Unfortunately, the bulk of plastics used are thermoplastics (§4.1.1), and some of these start to soften at temperatures well below 100°C. Thermosets (§4.1.2), while

Figure 8.42 Tube cutting using an air plasma cutting torch. (Courtesy of Goodwin Air Plasma Ltd, Loughborough.)

they do not soften, cannot tolerate temperatures much in excess of 250°C, and it is therefore clear why any method used to machine plastics must involve minimal heating of the workpiece material.

This is achieved by using tooling that has very high rake and clearance angles (Fig. 8.19), a copious coolant supply, small depths of cut and low feed rates. Tools must be kept extremely sharp at all times to prevent cutting forces, and hence temperatures, from increasing above the absolute minimum.

Another property of most plastics should not be overlooked, and that is their tendency to absorb water and swell up as a result. Because most coolants are aqueous solutions (water based), prolonged exposure to cutting fluids could result in a component machined to the correct size being undersize when it dries out. Fortunately, most plastics are not machined to such fine tolerances for this to be a serious problem, but if high precision is a requirement then grinding using a silicon carbide, coarse grit wheel with compressed air cooling may be a solution.

8.11 Summary of the principal characteristics, advantages and disadvantages of metal removal processes

Process	Characteristics	Advantages	Disadvantages
Lathes: centre, capstan, turret, single and multispindle automatics	Workpiece rotates and the cutting tool is slowly traversed in the desired direction – usually either parallel to the workpiece axis of rotation (turning) or at right angles to it (facing and parting off)	Ideal for circular parts, and for those with large length/diameter ratio Centre lathes are highly versatile Automatics offer high productivity with minimal manning	Limited level of accuracy and surface finish possible, and often requires additional operations such as grinding Highly skilled labour needed to exploit centre lathe's potential fully Automatics are expensive to tool and set up – a multispindle auto can take up to a day
Shaping and planing machines	Cutting tool reciprocates while workpiece is slowly traversed below it (shaping). In planing the tool is traversed across the reciprocating workpiece	Best suited for creating flat surfaces Planing machines capable of handling extremely long (fabricated) workpieces	Productivity restricted owing to cutting on forward stroke only, although rapid return mechanism helps Some planing machines have the facility to permit cutting on both strokes Shaping machines very limited in length of job possible because of the maximum practical stroke of the ram
Milling machines: vertical and horizontal	Workpiece is slowly traversed under a rotating multiedge cutting tool	Primarily designed to produce flat surfaces, although curved surfaces can be produced using form tools Flexibility can be greatly enhanced by use of accessories such as a dividing head (DeGarmo et al. 1988)	Size of component is limited, and skilled labour is needed to achieve the full potential of these machines Cutting tools can be expensive to buy and maintain
Drilling and boring machines	Workpiece stationary and rotating cutting tool fed into it either to drill or bore a hole or to create a flat surface	Drilling and tapping can be carried out on the same drilling machine Precision facing can be performed on the same machines as are used for large hole boring Multiedge facing cutters as used in milling can be used in borers for quicker machining of flat faces	Mechanical handling of large parts can be a problem Semiskilled labour is adequate for drilling, but skilled staff are needed for precision machining on borers

Process	Characteristics	Advantages	Disadvantages
Abrasive machines: grinders, honing and superfinishing machines, lapping machines	All these machines use some form of abrasive – either a grinding wheel, honing sticks or abrasive lapping paste – as the cutting "tool" Either workpiece or tool is the prime moving member, depending upon the type of machine and its design	*This is the only machine tool family that can cut hardened components* Very high levels of accuracy and surface finish are readily achieved Almost any shape can be precision ground, including threads and gears	Most processes are slow and require highly skilled labour, particularly grinding Parts are usually at their maximum value when ready for grinding so errors can be expensive Honing cannot correct hole location errors
Mechanical NTM processes Ultrasonic machining	Grits suspended in a fine abrasive slurry are hammered into the workpiece surface by a form tool resonating at 20 kHz. This causes the surface to be eroded/chipped away to leave the required cavity	No thermal stresses set up in workpiece Can cut non-conductive, hard materials Ideal for delicate components as cutting force is low No burrs on finished holes Holes complete in one "pass"	Only brittle materials can be cut Cutting rates are slow Significant tool wear
Water jet machining	Water at extremely high pressure (3000–4000 bar) is fired at the surface to be cut. For cutting metals, glass, etc. an abrasive powder is added to the jet stream to increase its penetrative power	Materials of any hardness can be cut Cutting can be in any direction – even three-dimensional is possible with computer control of nozzle motion No deburring is required No thermal stresses in the workpiece	Limited nozzle life, particularly when abrasive is used High capital cost of equipment (£500 000 is a typical figure) Because of supersonic nozzle velocities this is a very noisy process
Electrochemical machining (ECM) processes cutting, grinding, deburring	Reverse of electroplating and similar to electro-polishing, i.e. metal is stripped from work surface by electrolytic erosion	Any hardness of material can be cut No physical contact with workpiece so no cutting or thermal stresses generated – good for delicate parts Long tool life No edge burrs Holes complete in one "pass" Grinding wheel wear reduced by 90 per cent when electrochemical grinding	Workpiece must be electrically conductive Capital equipment expensive, particularly the electrolyte pumping and cleansing systems Form tooling costs can be high, particularly where 360° cutting is required
Chemical machining chemical etching, blanking and milling	Controlled surface corrosion of unmasked material. With chemical blanking corrosion continues right through the full thickness of the workpiece	Low capital cost once basic equipment is installed Many work surfaces can be "machined" simultaneously No cutting forces so no	Undercutting of masked areas a problem Sharp corners not possible Workpiece material must be homogeneous Disposal of chemical

Process	Characteristics	Advantages	Disadvantages
		distortion or thermal stress problems –good for delicate parts No edge burring	etchant expensive Operators require special protection against possible chemical spillage
Thermal NTM processes Electro-discharge machining (EDM)	Metal surface is spark eroded (blasted away) by a series of electrical discharges, leaving the required cavity Can be either used for die sinking shapes or for profile cutting via wire erosion	Materials of any hardness can be cut High accuracy and good surface finish are possible with suitable adjustment of process control parameters. No cutting forces involved Intricate-shaped cavities can be cut with modest tooling costs Holes completed in one "pass"	Limited to electrically conductive materials. A slow process, particularly if good surface finish and high accuracy are required Dielectric vapour can be dangerous HAZ near cutting edges Die sinking tool life is limited and wire must be scrapped after one pass in wire eroding
Electron beam machining (EBM)	Hole drilling and profile cutting by vaporizing the workpiece material with a high-energy beam	Any material can be cut, and as no cutting forces are involved, no stresses are imposed on the part being machined Exceptional drilling speeds possible, and with highly accurate position and form Extremely small kerf width when profiling, so little material is wasted	Skilled labour required A suitable backing material must be used Maximum thickness that can be cut is limited to about 10 mm and drilling length/diameter ratio up to 15:1 Work area must be under vacuum, which takes time to establish Capital equipment cost is high
Laser beam cutting	Another form of high-energy beam used to melt/vaporize the surface of the workpiece when drilling, profile cutting and surface hardening	Virtually any material can be cut Two-dimensional and three-dimensional shapes can be cut providing computer control of the laser beam nozzle is used Holes having a length/diameter ratio of up to 100:1 are possible High cutting speeds possible with thin materials, but much slower as material thickness is increased No consumable tooling needed The working area need not be under vacuum as is required with EBM cutting	Holes bigger than 1 mm diameter difficult to drill HAZ formed around cut surface edges Materials with a highly reflective surface difficult to machine The accuracy of the more powerful CO_2 laser is poor compared with the lower powered Nd:YAG type, but as a result cutting speeds are much slower with Nd:YAG lasers
Plasma arc cutting	Yet another form of high-energy beam, comprising an extremely high-	Any material can be cut, up to about 150 mm thick Much faster than the	Quite large HAZ around the cut edges Sharp corners not

Process	Characteristics	Advantages	Disadvantages
	temperature ionized gas stream, which is used to melt the surface of a workpiece when profile cutting	usual alternative of oxy-fuel cutting Cutting can be in any direction	possible, and the cut edge is usually slightly bevelled. Electrode life is short – typically 2–10 hours cutting depending upon the number of starts Kerf width is up to twice that of oxy-fuel cutting Noisy process

Bibliography

Anonymous 1985. Using ECM to produce high quality parts economically. *TechCommentary*, **2**(6), 1–5.

Benedict, G. F. 1987. Nontraditional manufacturing processes. New York: Marcel Dekker.

Boothroyd, G. 1965. *Fundamentals of metal machining*. London: Edward Arnold.

Castrol 1990. *Talking about health and safety – lubricants and allied products*, 3rd edn. Swindon: Castrol (UK).

Castrol 1991. *Talking about cutting fluids*, 9th edn. Swindon: Castrol (UK).

David Brown Special Products Undated. *David Brown basic gear book*. Huddersfield: David Brown Special Products.

DeGarmo, E. P., J. T. Black & R. A. Kohser 1988 *Materials and processes in manufacturing*, 7th edn. New York: Macmillan.

Messer Griesheim Undated. CNC *electron beam drilling machines*, publication no. 47.2010e. Cramlington, Northumberland: Messer Griesheim.

Goodwin, D. Undated. *An introduction to the air plasma cutting process*. Loughborough: Goodwin Air Plasma.

Kalpakjian, S. 1984. *Manufacturing processes for engineering materials*. Reading, MA: Addison-Wesley.

Lindberg, R. A. 1983. *Processes and materials of manufacture*. Newton, MA: Allyn & Bacon.

Lissaman A. J. & S. J. Martin 1982. Principles of engineering production. London: Hodder & Stoughton.

McGeough, J. A. 1974. *Principles of electrochemical machining*. London: Chapman & Hall.

Sandvik Coromant 1994. *Modern metal cutting*. Sandviken, Sweden: Sandvik Coromant.

Case study

A pump company produces a range of pumps fitted with low-carbon steel shafts. As part of an efficiency exercise the production planning department was asked to determine the effects of increasing by 50 per cent either the cutting speeds or feed rates currently used when turning these shafts.

Illustrate how such assessment calculations are carried out and draw relevant conclusions, basing your example upon the following parameters and using, where necessary, data contained in Tables 8.1 and 8.2.

Initial material diameter = 60 mm, depth of first roughing cut = 5 mm, feed rate = 0.4 mm/rev, spindle speed = 530 rev/min, length of cut = 150 mm. The lathe used is fitted with a 25 HP motor; assume that the overall efficiency of the machine tool is 70 per cent.

The Taylor tool life equation constants applicable to the workpiece/cutting tool material combination and the above cutting parameters are $C = 158$ and $n = 0.117$.

While steel shafts are currently used in all the company's pumps, for technical reasons it is proposed that one particular pump model should in future be fitted with a cast iron shaft. The production planning department was therefore also asked to determine the effects on machining times and tool life of such a material change. For comparative purposes, again consider 60 mm diameter material and depth of cut = 5 mm, but a feed rate of only 0.8 mm/rev and linear cutting velocity of 80 m/min – typical figures for dry cutting cast iron using carbide tooling.

The Taylor tool life equation constants applicable to this workpiece/cutting tool material combination and the above cutting parameters are $C = 153$ and $n = 0.176$.

Solution

The algebraic relationships used in this case study are all to be found in sections 8.4, 8.5 and 8.6.

Low-carbon steel shafts

The current cutting conditions are:
- linear cutting velocity $(V_c) = \pi DN = (\pi \times 60 \times 530)/1000 = 100$ m/min
- depth of cut $(a) = 5$ mm
- feed rate $(f) = 0.4$ mm/rev

- power currently being used $= \dfrac{\left(af V_c k_{c \text{ modified}}\right) \times 1.25}{\eta \times 1000 \times 60}$

$$= \dfrac{\left(5 \times 0.4 \times 100 \times k_{c \text{ modified}}\right) \times 1.25}{0.7 \times 1000 \times 60}$$

- k_c (from Table 8.1) for low-carbon steel = 2200 N/mm^2
- correction coefficient (Table 8.2) = 1.00 for a feed rate of 0.4 mm/rev.

Thus, actual lathe motor power currently used

$$= \dfrac{5 \times 0.4 \times 100 \times 2200 \times 1.25}{42\,000} = 13.1 \text{ kW} = 17.6 \text{ HP}$$

If spindle speed is increased by 50 per cent (linear cutting velocity then 150 m/min), motor power would also increase by 50 per cent as power is directly proportional to cutting speed, i.e. to 26.4 HP.

If feed rate is increased by 50 per cent, this affects the k_c correction coefficient, and for 0.6 mm/rev = 0.89 (Table 8.2). Thus, k_c becomes $2200 \times 0.89 = 1958$.

Therefore increasing feed rate by 50 per cent requires motor power of:

$$= \dfrac{5 \times 0.6 \times 100 \times 1958 \times 1.25}{42\,000} = 17.5 \text{ kW} = 23.4 \text{ HP}$$

Tool life under current cutting conditions,

$$T = \left(\frac{C}{V_c}\right)^{\frac{1}{n}} = \left(\frac{158}{100}\right)^{\frac{1}{0.117}} = 50 \text{ minutes}$$

Increasing feed rate has minimal adverse effect on tool life, but increasing spindle speed by 50 per cent to 150 m/min results in a tool life,

$$T = \left(\frac{158}{150}\right)^{\frac{1}{0.117}} = 1.6 \text{ minutes}$$

The time to machine one 150 mm length, when cutting at the current spindle speed of 530 rev/min and 0.4 mm/rev is 0.71 minutes. So the equivalent of 70 roughing cuts can be made before tool life is exceeded. If spindle speed is increased by 50 per cent to 795 rev/min, then although it takes less time to execute each cut the resultant dramatic reduction in tool life means that only three roughing cuts per tool life can be expected!

Cast iron shaft

$$\text{Power required} = \frac{\left(5 \times 0.8 \times 80 \times k_{c \text{ modified}}\right) \times 1.25}{0.7 \times 1000 \times 60}$$

Also, k_c (from Table 8.1) for cast iron is 1300 N/mm^2 and the correction coefficient (Table 8.2) is 0.82 for a feed rate of 0.8 mm/rev. Thus, $k_{c \text{ modified}} = 1066$.

$$\therefore \text{ motor power required} = \frac{5 \times 0.8 \times 80 \times 1066 \times 1.25}{42\,000} = 10.2\,\text{kW} = 13.6\,\text{HP}$$

Time to machine a 150 mm length is 0.44 minutes at $(80/100 \times 530)$ rev/min, i.e. 424 rev/min.

$$\text{Tool life} = \left(\frac{C}{V_c}\right)^{\frac{1}{n}} = \left(\frac{153}{80}\right)^{\frac{1}{0.176}} = 40 \text{ minutes}$$

Conclusions

It is clear from these calculations that increasing feed rate by 50 per cent increases production output by a similar amount, and without a significant reduction in tool life. Motor power required also increases by 50 per cent, but this is still within the capabilities of the lathe used.

Increasing cutting speed by 50 per cent is quite another matter in that, while increasing production output by 50 per cent per unit time, tool life is drastically reduced from 50 minutes to just 1.6 minutes. Thus, tooling costs will increase dramatically because of both the much greater number of tools required and the increased non-cutting time (downtime) resulting from the more frequent tool changes necessary. Furthermore, the lathe's motor power is not quite adequate.

Changing to a cast iron shaft from a steel one results in a slightly shorter tool life (40 minutes compared with 50 minutes) even though, because of the higher feed rate (0.8 mm/rev), cutting time is reduced from 0.71 minutes to 0.44 minutes, i.e. an increase in productivity of nearly 40 per cent. Furthermore, this is achieved with almost 17 per cent less power than is needed when machining steel shafts.

Questions

8.1 What is meant by the term surface generation? Illustrate your answer by defining the essential movements of cutter and workpiece to cut cylindrical, flat and conical shapes on a centre lathe.

8.2 What advantages and limitations does the use of form tooling have over pure generation when producing a complex profile?

8.3 Why has the use of copying attachments as a means of repetitive shape production been in serious decline in recent years?

8.4 Why are the lathe family of machine tools more suited to the production of cylindrical shapes than flat ones?

8.5 What are the two major limitations inherent in the design of shaping machines? What does the machine tool industry offer as an alternative?

8.6 Conventional (not CNC) machine tools have gear boxes whose gear ratios are in geometric rather than arithmetic progression. Why do you think this is?

8.7 Precise alignment of the principal axes of any machine tool is essential if accurate work is to be produced. A universal milling machine that has been previously used in its horizontal milling mode is reset to operate as a vertical miller. Unfortunately, the machine tool setter does not align the vertical axis (the rotational axis of the cutter spindle) exactly normal (90°) to the horizontal bed of the machine. Instead it is set at an angle of 89 57'15". If the operator now uses this machine to bore a vertical hole in a workpiece mounted on the machine's bed, show that the resulting shape will be an ellipse.

8.8 Why are horizontal boring machines so useful in the manufacture of gearbox casings?

8.9 Suggest at least three reasons why a production engineer might choose to remove metal by an abrasive cutting process rather than by using a single- or multiedge cutter.

8.10 Suggest three design features of grinding machine tools that enable them to produce components to extremely small dimensional limits. If they are located among other types of machine tools, what precautions must be taken and why?

8.11 Why does a lap not wear even though it is softer than the component being cut?

8.12 Honing can establish, with great precision, certain workpiece dimensions but not others. What dimensions can it not influence, and why?

8.13 Give reasons why oblique cutting is preferable to orthogonal cutting.

8.14 What is the rake angle of a single-edge cutting tool, and how does it affect the shearing action when metal cutting? How is it affected by the approach angle of an oblique cutting tool?

8.15 What are the main differences between rough cutting and finish cutting?

8.16 Discuss the statement "a grinding wheel is similar to a milling cutter", by comparing and contrasting the cutting action of a grinding wheel with that of a multiedge milling cutter.

8.17 Describe the self-sharpening action of a grinding wheel when it is cutting correctly. What is meant by the statement "the wheel is cutting soft"?

8.18 What is grinding wheel dressing? When and why is it carried out?

8.19 Why is specific cutting pressure a useful parameter in assessing the cutting characteristics of a given material?

8.20 Before the advent on modern CNC machine tools, a machine tool builder produced a lathe largely for facing operations, offering an arrangement allowing for constant linear cutting velocity rather than the normal constant rotational speed. Suggest some of the advantages offered by this machine, and calculate the cutting time for facing between 500 mm and 50 mm diameter, starting at a rotational speed of N rev/min. Compare this with constant speed machining at N rev/min. Take feed rate = 0.15 mm/rev and N = 180 rev/min.

8.21 What effect does depth of cut and feed rate have on the life of a cutting tool?

8.22 For a particular cutting operation the required cutting life of a single-edge tool is 20 minutes. The operator has a choice of tools, one being made from HSS and the other from tungsten carbide. What difference in linear cutting velocity will there be between these two tools if the Taylor constants for HSS in this case are $C = 250$, $n = 0.125$, and for tungsten carbide $C = 650$, $n = 0.29$?

8.23 What are the essential properties of any cutting tool material? What is meant by the term "hot hardness"?

8.24 What is the difference between a grinding wheel's grade and its structure?

8.25 What are the main objectives of a cutting fluid and why are most modern coolants water based?

8.26 What advantages do non-traditional machining processes such as EDM, ECM and laser beam machining have compared with conventional metal cutting? Why do you think that manufacturing industry has been slow in adopting such advanced machining processes?

8.27 What are the main differences between EDM and ECM? Give an indication of typical metal removal rates and surface finishes achievable, and discuss the main advantages and disadvantages of each process.

8.28 Is ultrasonic machining really a chipless machining process? Why is it only practicable to cut brittle materials with this process?

8.29 Why is chemical etching so attractive to printed circuit board manufacturers?

8.30 Compare and contrast laser and electron beam machining. State clearly the limitations of each process, and suggest at least one area of industrial application for electron beam drilling.

8.31 Even though plasma arc cutting creates a much larger kerf than conventional oxy-fuel cutting, and its electrode life is short, why is it still attractive as a profile cutting process?

8.32 What is the prime requirement when machining any plastic? What steps are usually taken to ensure that this consideration is met as far as possible?

Chapter Nine

Computer numerical control of machine tools

After reading this chapter you should understand:

(a) what CNC is and how it differs from NC
(b) the three constituent parts of any CNC system
(c) how part programs are constructed and then adapted to run on a specific machine tool
(d) the convention by which CNC machine tool axes are defined
(e) the difference between absolute and incremental programming
(f) why canned cycles and G and M codes are used
(g) the many tasks and features of a modern machine control unit
(h) the four modes of machine tool axis control
(i) the difference between tooling offset and compensation
(j) the fundamental design differences between conventional and CNC machine tools
(k) why touch trigger probing should be used on CNC machine tools.
(l) the benefits of adaptive control
(m) the difference between a flexible manufacturing cell and system.

Computers have changed almost every aspect of life over the past 25 years, and the design of machine tools has not been exempt from this technological revolution.

While it is still possible to buy conventional machine tools of the type discussed in Chapter 8 (§8.2), most machine tools sold today have some degree of computer control. For example, Cincinnati Milacron, one of the world's largest machine tool builders and leaders in the milling machine market for most of this century, has sold virtually no conventional milling machines for approximately 10 years (Cincinnati Milacron, Birmingham, personal communication).

The purpose of this chapter is therefore to provide the reader who already has a basic understanding of conventional machine tools with an insight into the underlying principles of computer numerically controlled (CNC) machine tools.

9.1 Machine tool control

The operation of any mechanical device must be controlled in some way; control may be either manual or automatic in response to a preprogrammed sequence of instructions. In the case of machine tools, such instructions may be in the form of cyclically operating cams, levers, microswitch end-stops, etc. Although high rates of production can be achieved with this form of rigid automatic control (§10.1.3), it is inflexible, and prolonged machine set-up times are needed to change from one product to another. To increase flexibility and to reduce set-up times, computer programs are increasingly being used as the means of issuing the required control instructions to machine tools.

Other benefits of replacing manual control with computer control are summarized in §9.4.

9.2 Computer numerical control (CNC)

The three basic constituent parts of any CNC system are (Fig. 9.1) a part program, an electronic controller – usually referred to as a machine control unit (MCU) – and the machine tool itself. The key features of each of these three elements are now described.

Note   Computer numerical control is *not* a manufacturing process but a method of electronic machine control.

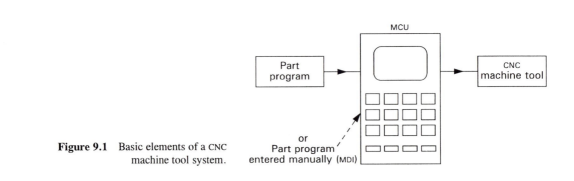

Figure 9.1   Basic elements of a CNC machine tool system.

9.2.1 The part program

The part program is the source of all instructions required by a machine tool for it to produce the required component. In other words, it replaces the input provided by a skilled machinist when operating a conventional machine tool.

Note   *Part program* is perhaps an unfortunate term as the uninitiated might interpret it to mean a part or portion of a complete component program. Although component program would be a more unambiguous description, part program is the standard term used in industry and is therefore the one used in this book.

To appreciate how part programs are usually constructed, it is helpful to consider in more detail how a skilled operator, presented with the drawing of a new component, produces the required item on a conventional machine tool. First, the drawing is carefully examined to establish the overall shape and dimensions of the part, the accuracy required and the material from which it is to be made. This information then enables the operator to select the size of raw material, the most suitable tooling and the speeds, feeds and depths of cut that should be used.

To enable a CNC part program to emulate this skilled operator's mental processes, CNC programmers have found that it is most convenient to divide the programming structure into two sections. One section contains the dimensional information that geometrically describes the part required, called the *geometrical data*, and the other section contains all the data necessary for the machine tool to select the required tooling in the correct sequence, cut at the correct speeds, feeds and depths of cut, and even turn the coolant on and off automatically. This portion of the program contains what is termed the *technological data*.

As this approach is broadly the same as that used by production planning engineers when they plan the production procedures to be used by conventional machine tool operators (§8.2), it is not surprising that the most successful CNC programmers are usually the most experienced production planning engineers. They write using a user-friendly alphanumeric language that is not dependent upon the specific machine tool upon which it is to be used. The reason for this is that there may be a number of different machine tools on which a given component can be made, but unfortunately different machine tool builders use different program protocols (language conventions). So, to avoid wasting time rewriting a given part program for every machine tool on which it might possibly be used, the program is written in one of a number of variants of the basic *automatically programmed tool* (APT) language. It is not an aim of this chapter to teach part programming, but interested readers will find examples of such programming languages in texts written specifically for that purpose (Gibbs 1987, Thyer 1991).

The general part program is next passed through a special computer program, termed a processor, which converts the alphanumeric program into computer machine code, checks for programming inconsistencies, expands shorthand instructions into full individual computer instructions and calculates the detailed path that the cutting tool's centre-line must follow to ensure that the required shape is produced (§9.2.2.4). However this processor's output is still not machine tool specific.

Finally, to turn the program into one that can be understood and acted upon by a given machine tool's MCU (§9.2.2), it is put through another computer program, called a post-processor. Each machine tool builder should provide appropriate post-processor software at the time of supplying the machine, although more experienced companies often write their own as required (Fig. 9.2).

It is also possible, with modern MCUs, for skilled operators to program a part actually at the machine tool, using the MCU's keyboard and screen (VDU). This is discussed in more detail in section 9.2.2.2.

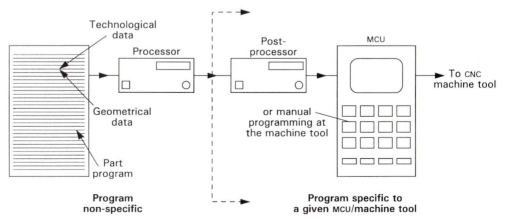

Figure 9.2   Steps in producing a part program for a specific machine tool.

9.2.1.1 Axis convention

In mathematics a universally accepted system of sign and axis conventions is used. For example, when plotting a graph it is usual to define the vertical axis as y and the horizontal axis as x. It is equally important, when entering the geometrical (shape) data into a CNC program, that a similar unambiguous system exists for each axis of the machine tool with respect to the workpiece.

While there are minor differences in axis conventions used in the CNC machine tool industry, the vast majority use the same conventions as used in graphical mathematics: for machines with up to three orthogonal axes x, y and z are used to describe each linear axis, with A, B and C being used for rotary motion about each of these three axes respectively (Fig. 9.3). This convention conforms to ISO Recommendation R841 and BS 3635: Part 1.

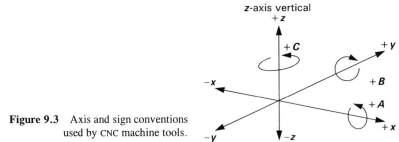

Figure 9.3 Axis and sign conventions used by CNC machine tools.

The axis of the machine's main spindle is always designated the z-axis, irrespective of whether this is the axis about which the cutting tool or workpiece rotates.

Motion in the z-axis is always considered positive in the direction that takes the workpiece away from the tool holder. This is because, if the programmer accidentally forgets to enter a minus sign in front of a co-ordinate, the cutting tool and workpiece move further apart rather than closer together, as mathematically the absence of a sign is taken as a plus ($5.5000 = +5.5000$). This is an important anti-crash safety feature. If the z-axis is known, the x- and y-axes can be established from the right-hand rule in the following manner (Fig. 9.4).

Figure 9.4 "Right-hand rule" for machine tool axis designation. (Courtesy of H. Kief, Michelstadt, Germany.)

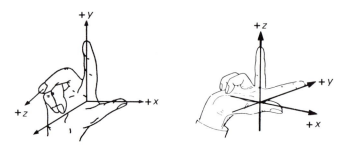

Axis designation for machine tools with horizontal z-axis

Axis designation for machine tools with vertical z-axis

244

Imagine the middle finger of your right hand to be in the main spindle of the machine (the z-axis), with the positive direction being, for safety reasons, in the direction of tool retraction. Next turn your hand so that the thumb points in the plane of the longer of the two remaining principal axes – this is the x-axis. Your index finger is now pointing in the plane of the y-axis.

Exercise   Figure 9.4 illustrates the right-hand rule applied to machines having either a horizontal spindle axis (as with a lathe) or a vertical spindle axis (as with a vertical milling machine). Sketch the outline of a simple lathe showing the x-, y- and z-axes, not forgetting to indicate which direction is positive and which is negative. What would cause motion in the y-axis?

Positive rotation in the rotary axes A, B and C is counter-clockwise when viewed from the + direction of its corresponding linear axis: x, y and z (Fig. 9.3).

9.2.1.2 Absolute and incremental programming

The geometrical data contained in a part program, which defines the physical shape and size of the part to be made, is normally specified by a series of cartesian co-ordinate points. They may be expressed in two ways – as absolute or incremental co-ordinates.

Note   While it is possible to program geometrical data using polar co-ordinates, CNC programmers tend to favour cartesian co-ordinates (Fig. 9.5). Some programs are best written using a combination of both, but this presents no problems for modern MCUs.

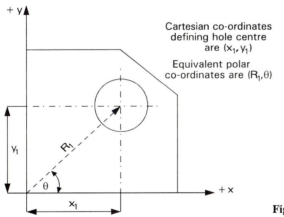

Cartesian co-ordinates defining hole centre are (x_1, y_1)

Equivalent polar co-ordinates are (R_1, θ)

Figure 9.5   Cartesian and polar co-ordinates.

All CNC programs define their geometrical data relative to some suitable datum point, and this is the starting point for either method of specifying a sequence of co-ordinate points. Traditionally programmers choose the bottom left-hand corner of the component when programming milled or drilled parts (Figs 9.6a and 9.7), i.e. the x and $y = 0$ point, and when programming turned parts the centre of the end of the workpiece furthest from the chuck is normally selected, i.e. the z and $x = 0$ point (Fig. 9.6b).

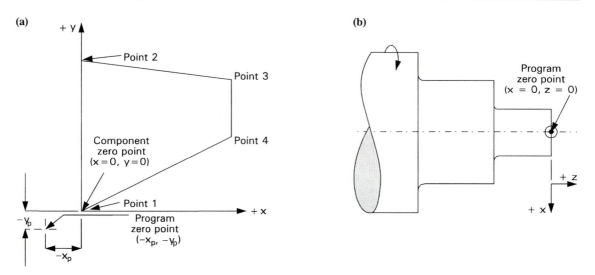

Figure 9.6 Commonly used co-ordinate programming datum points.

Note Component datum zero point need not necessarily be the program zero point, although it frequently is. Figure 9.6a illustrates how simple it is to relate co-ordinates from one datum point to another that is offset by, say, $-X_p$ and $-Y_p$.

Defining points using absolute co-ordinates involves specifying each point from the zero datum point (Fig. 9.7a), but in incremental dimensioning (also called chain dimensioning) each point is expressed as the path differential from the preceding point (Fig. 9.7b).

The principal advantage of programming using absolute co-ordinates is that, unlike incremental dimensioning, any point can be readily changed without affecting subsequent dimensions. Also, re-entering a program after an unexpected interruption for example caused by a power loss, is much easier when programming is based upon absolute dimensioning.

However, incremental programming does have certain advantages, such as the simplicity of copying and transferring specific geometries such as drilling patterns, chamfers, fillets and

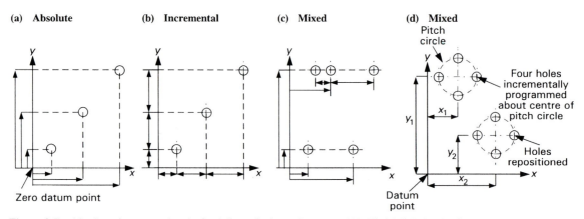

Figure 9.7 Absolute, incremental and mixed dimensioning. (Courtesy of H. Kief, Michelstadt, Germany.)

246

milling cycles when their dimensions are incrementally specified. For example, if a pattern of holes is to be repositioned on a component it is only necessary to alter the x- and y-co-ordinates of the pitch circle from which the hole pattern is dimensioned from x_1, y_1 to x_2, y_2 (Fig. 9.7d).

Also, the sum of both the x and y incremental dimensions must always equal zero when returning to the start position at the end of a complete cycle (see the example below). This makes it easy to check a program's dimensional accuracy, but is not possible, of course, when absolute dimensioning is used or when incremental and absolute dimensioning are mixed.

Both forms of point definition are acceptable to modern MCUs, which may be switched between each format at will without losing the zero datum point of the co-ordinate system in use.

As stated previously, the intention of this chapter is not to teach part programming, but to illustrate clearly the difference between absolute and incremental dimensioning, the following simple example is included.

Example

A series of seven holes are to be drilled in a component to the pattern shown in Figure 9.8.

Table 9.1 lists the dimensions of the seven holes, using both absolute and incremental methods. Note that the incremental co-ordinates summate to zero, suggesting that the programming is correct.

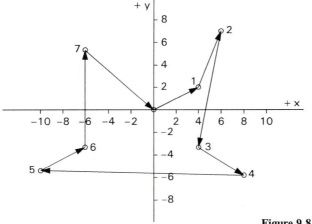

Figure 9.8   Required hole pattern.

Table 9.1

Hole no.	Absolute dimensioning		Incremental programming	
	x	y	x	y
1	4	2	+4	+2
2	6	7	+2	+5
3	4	−3	−2	−10
4	8	−6	+4	−3
5	−10	−5	−18	+1
6	−6	−3	+4	+2
7	−6	+5	0	+8
0	0	0	+6	−5
			$\Sigma = 0$	$\Sigma = 0$

9.2.1.3 Canned cycles

Every machine tool has certain machining cycles that are repeated continuously. They are geometric sequence programs with only minor, if any, parameter changes ever being made by the programmer. To simplify programming, such repetitive routines are permanently stored in the MCU, and are called up at will by the appropriate programming command. Such stored programming sequences are referred to as canned cycles, and range from simple coolant on/coolant off instructions to more complex sequences for drilling and tapping of holes.

Many of these canned operations/cycles have been standardized by the machine tool industry. They have been divided into two broad categories called "preparatory functions" and "miscellaneous functions", the former being called G functions and the latter being known as M functions.

G functions are related to how the machine tool treats the *geometric* data supplied, such as whether dimensions are in inches or metric units (G70 and G71), whether absolute or incremental dimensioning is being used (G90 and G91), program zero reference point shift (G92), etc.

M functions are more concerned with controlling the machine tool's *mechanical* functions such as turning coolant on and off (M07 and M08), switching the main spindle on and off (M03 and M05), program stop (M00), initiating automatic tool change (M06), etc.

Although certain basic G and M codes are now almost universal (Kief & Waters 1992), many other G and M functions unfortunately differ from one machine type and builder to another, and great care must therefore be exercised to ensure that the correct codes are used.

While G and M codes are a useful short-hand and simplify programming in the planning office, they can be even more valuable when manually programming at the machine tool, as they both speed up programming and reduce the risk of programming errors. Modern MCUs even permit the machine tool user to program a few special macros (canned cycles), some G and M numbers being left unassigned for this purpose.

Exercise Can you see the similarity between canned cycles used in CNC machine tool programming and a similar procedure used by computer programmers? If not, briefly examine any basic textbook on computer programming and read the section dealing with macros.

9.2.1.4 Modal and non-modal functions

Certain G and M functions, once included in a part program, continue to operate until cancelled or countermanded by a new instruction. These are called modal functions, a good example being G71, which instructs the program to treat all dimensions as metric. Those functions that, once operated upon, are then automatically cancelled and must be re-entered in the program when next required are termed non-modal functions. A typical example of this sort of function is M00, which instructs the machine tool to stop and turn off both spindle and coolant supply.

9.2.2 Machine control units

Having produced a suitable part program, it is vital that the machine tool on which the component is to be made is precisely controlled to ensure that it executes all the program's instructions accurately and in the correct sequence. This is the job of the sophisticated electronic controller attached to all CNC machine tools (Fig. 9.9), and which is normally referred to as the machine control unit (MCU).

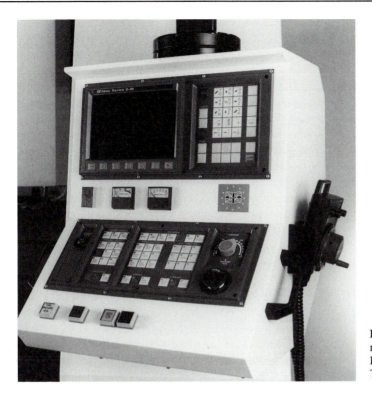

Figure 9.9 Typical CNC machine tool MCU. (Courtesy of Kryle Machine Tools, Stoke-on-Trent.)

9.2.2.1 The principal functions of an MCU

The MCU takes a part program, irrespective of its source, and translates its contents into electrical signals, which it then sends out as movement instructions to the various machine tool elements so that the required sequence of events is executed. The two main elements of any CNC machine tool that are controlled in this way are the main spindle motor and the axes drive motors. Axis drive motor control can be complex as all axes must be continuously regulated both individually and in relation to each other, because it is the precise relative movement of the different axes that generates the required two-dimensional or three-dimensional shape. This topic is discussed in more detail in §9.2.2.3.

Many other functions are also performed by the MCU, such as complex calculations related to tooling offsets and cutter path correction (§9.2.2.4), automatic tool changing, cutting tool life monitoring (§9.2.2.5), automatic tool setting and in-cycle gauging (§9.2.2.6) and coolant control.

One of the most important facilities available on modern controllers is their ability to detect system faults. This interactive self-diagnostics capability is becoming ever more sophisticated, and it is now possible for MCUs not only to detect faults, but also to pinpoint the cause(s) and recommend the most appropriate corrective action. This has made the problems of maintaining complex electromechanical CNC systems much less costly, particularly for smaller companies, which previously had to call upon expensive external service organizations for even the most minor system faults.

9.2.2.2 Inputting programs into an MCU

The first numerically controlled (NC) machine tool controllers had no computing capability and received their instructions from part programs stored on rolls of paper tape with holes punched in them to a predefined code (Keif & Waters 1992). While this form of program storage can still be found on a few machines, paper tape has largely been superseded by magnetic recording media such as magnetic tape or floppy disk.

Increasingly machine tools are having part programs downloaded on demand directly to their MCU from a central host computer – referred to as direct numerical control (DNC). However, this has only become possible because modern controllers are now fitted with significant memory capacity. Indeed, such controllers are capable of storing a number of programs that can be called up at will by the operator – a useful backup in the event of host computer failure. Programs can also usually be loaded from a portable PC directly into the MCU at the machine tool.

Note DNC involves much more than just the downloading of programs to individual CNCs from a host computer. It also involves many other aspects of interactive system communication (Keif & Waters 1992), a detailed study of which is beyond the scope of this chapter.

Another major difference between early machine tool controllers and modern ones is that the latter now have powerful computing ability. Thus it can be said that:

$$CNC = NC + \text{program storage capacity and computing capability}$$

When the programming office is unable to respond to an urgent programming request, it is also possible with modern MCUs offering major computing capability for experienced operators to input programs directly into the MCU using its alphanumeric keyboard and VDU. This is termed manual data input (MDI). While MDI eliminates the need for computer processing and post-processing of the program (Fig. 9.2), this part program is nevertheless machine tool specific, i.e. it can only be used on CNC machine tools with the same type of MCU. Therefore, un-

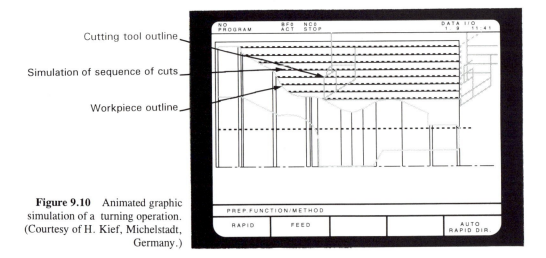

Figure 9.10 Animated graphic simulation of a turning operation. (Courtesy of H. Kief, Michelstadt, Germany.)

less the part is simple and/or extremely urgent, programming should be carried out away from the shop floor in the planning office. Consistent planning techniques are then assured, and the program can be proof tested before it is used to cut metal for the first time.

Most controllers also offer program proof testing, by running the program before metal cutting, and displaying cutter and workpiece movements as an animated graphical simulation on the VDU (Fig. 9.10).

Unfortunately only the most sophisticated MCUs take account of any clamping devices being used to hold the workpiece – the cause of many a cutter breakage. This is also a potential problem with programs written in the planning office, of course, and such factors must therefore be taken into account by the skilled operator.

9.2.2.3 The four modes of axis control in CNC machine tools

(a) *Point-to-point control* (Fig. 9.11a). The example depicted in Figure 9.8 involving the drilling of seven holes in a plate is a good example of point-to-point control. As long as the drill has been retracted clear of the work surface before it moves, the precise path taken in going from one hole to the next is irrelevant. Thus, the axes involved (in this example x and y) can be independently and rapidly traversed until each axis reaches its programmed target value. No cutting occurs during traverse, and cutting only commences once both target positions have been reached, the drill then being fed into the workpiece ($-z$) to the required depth. This control mode is simple and is typical of that offered by MCUs fitted to low-cost drilling and punching machines.

Fig 9.11a

Point-to-point

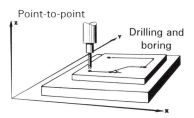

Drilling and boring

Fig. 9.11b

Point-to-point and straight line

Frame milling

Fig. 9.11c

Two-axis contouring with switchable plane

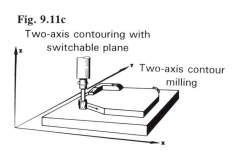

Two-axis contour milling

(b) *Straight line control* (Fig. 9.11b). All but the most basic two axis controllers permit each axis to be programmed individually to any desired feed rate. Each axis traverses completely independently of the other, but suitable selection of the feed rate for each axis allows the path taken by the cutter from one point to the next to be controlled. This is of value when milling a simple profile such as shown in Figure 9.11b, but as almost all controllers now offer at least full two-axis contouring (see c below), this control mode is now of minor interest and its area of operation tends to be limited to certain fields of robotic component handling.

(c) *Two-axis contouring with switchable axis control* (Fig. 9.11c). This method of two-axis spindle control differs from straight-line control in that the two axes are not moved independently. Instead the MCU controls both axes simultaneously at all times, adjusting the feed rate of each axis to provide the desired cutter path between one programmed point and the next. How this continuous axis interaction is actually achieved is explained in (d) below. This type of controller also offers the option of selecting which pair of axes are controlled (xy, xz or yz), as there are occasions when this can be beneficial and programming greatly simplified. However, the cost of full multiaxis MCUs has reduced to such an extent in relation to the cost of its associated machine tool that it is now rarely worth considering anything less than a full three-axis continuous path controller.

Fig. 9.11d

Three-axis continuous
path contouring

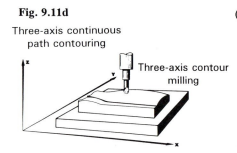

Three-axis contour
milling

(d) *Three-axis contouring with continuous path control* (Fig. 9.11d). Multiaxis continuous path control (contouring) systems provide accurate spindle positioning at any point in space because all axes are precisely controlled at all times, both individually and in relation to each other. This can be quite difficult to achieve in practice when three or more axes are involved, and the MCU achieves this task by using a built-in software package called an interpolator. This software co-ordinates the movement of each axis by calculating a series of three-axis target points along the intended path, and then controlling the relative motion of all axes to ensure that the target point for each axis is reached simultaneously. The greater the number of target points calculated, the more accurate will be the contour produced, but if three or more axes are involved (three-dimensional solid contouring), even with modern MCUs offering powerful computing capability, there is a risk that the time required to calculate each set of target points (x, y, z) can be longer than the machine tool takes to traverse between two such closely positioned sets of points. A compromise between profile accuracy and smooth machine tool control must therefore be adopted. Alternatively, programs calling for high-profile precision can be initially run remotely from the MCU, the large number of axis target points calculated being stored on disk and then downloaded to the MCU when required. This avoids the MCU having to calculate axis points during the actual cutting process. However, if the profile required can be easily defined mathematically, for example a straight line, circle, ellipse, parabola, etc., the MCU's computing speed is no longer a constraint, as interpolator software is usually capable of controlling the axes to produce such shapes without the need to calculate large numbers of multiaxis target points. Instead it employs linear, circular, parabolic or spline interpolation techniques, but the precise procedure involved is beyond the scope of this book.

9.2.2.4 Cutter offsets and compensation

During their useful life cutting tools vary in length and diameter from their nominal sizes owing to wear and as a result of resharpening. All CNC programs are based on nominal cutter sizes, so to take account of these unavoidable size variations MCUs are provided with features called cutter offset and cutter compensation. Thus, profiles, hole depths, etc. can continue to be produced correctly despite cutter size deviations from programmed nominal values.

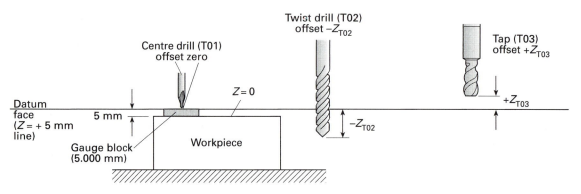

Figure 9.12   Tool length offsets for tools to drill and tap a hole.

COMPUTER NUMERICAL CONTROL (CNC)

Because the differences between cutter offset and cutter compensation are at times anything but obvious, and differ somewhat between turning and milling machines anyway, only a simplistic and generalized explanation of their differences will be given here.

When a number of tools of widely differing lengths and configurations are loaded into a machine's tool magazine, programming is much easier if their individual cutting edges are all referenced to a common datum – in milling or drilling often just clear of the top surface of the workpiece, say a plane at $z = +5$ mm (Fig. 9.12). The axes co-ordinates that define each tool's deviation from this common datum are termed tooling offsets, and are entered into the MCU by the operator at the time of initially setting up the machine's tooling. The controller then automatically adjusts for each tool's positional variations when executing the CNC program, and so avoiding the need to make changes to the basic program.

Another common form of tooling offset is related to profile milling. Programmers normally program profile geometry assuming a zero diameter cutter – very convenient as it means that co-ordinate points can be taken directly from the part drawing. This is called point programming. Because an MCU controls the path of the rotational axis of a cutting tool, in profile milling operations the path of the cutter's centre-line will differ from the profile actually cut by an amount equal to the cutter radius. To overcome this problem programmers specify in the technological portion of their programs the (nominal) cutter diameter to be used and, depending upon whether internal (pocketing) or external profiling is required, link this to a positive or negative offset instruction by the appropriate G code (§9.2.1.3). The MCU then automatically applies the necessary offset of the cutter centre-line equal to the nominal radius of the cutter used (Fig. 9.13).

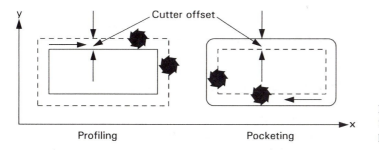

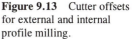

Figure 9.13 Cutter offsets for external and internal profile milling.

The other vital condition in profile milling is that the path followed by the rotational axis of the cutter *must be normal to the required profile at all times*. This is not as easy at is seems on curved surfaces, but fortunately is automatically catered for by the calculatory skills of modern MCUs.

The tool offset facility is also extremely useful when a limited range of sizes of the same component are required. Simply by changing offset values the required size differences can be achieved without the need for changes to the original part program.

Minor variations in cutter size from nominal (resulting mainly from wear) are also entered into the MCU's memory, but these are termed cutter compensations.

In practice, cutter compensation goes far beyond just compensating for minor differences in diameter and length, and is also used, for example, to compensate for the effects of nose radius found on most turning tools. However, it is sufficient for readers of this text to understand the basic principle of cutter offset and how it differs from cutter compensation.

Exercise   Draw the plan view of a square toolpost of the type found on conventional lathes, and which indexes about its geometrical centre. Show it holding four different forms of turning tool. Using the left-hand corner of the toolpost nearest to the workpiece as the tooling zero reference point, show the *z*- and *x*-axes tooling offsets applicable to each tool.

9.2.2.5   Cutting tool management

Another useful function of the modern MCU is its ability to both control automatic tool changing (see §9.2.3.4), assuming that the CNC machine tool in question is fitted with an auto-tool changer of course, and maintain a real time log of each cutting tool's use.

Each tool position in a machine's tool magazine is defined by a unique T number (T for Tool!). Providing each tool is where the MCU thinks it is then, as soon as the program calls for a given tool, the controller will initiate its automatic retrieval. In the case of CNC lathes tool retrieval normally involves the indexation of a multitool turret to bring the desired tool into the cutting position. The MCU also automatically calls up and applies any tool compensation and offset values that have been entered for the tool in question.

Modern controllers also have the ability to keep a running log of the total time that each cutter has been used. Providing maximum tool life values have been entered into the MCU, the controller will ensure that no tool is allowed to start a cut if, during that cut, it will exceed its scheduled tool life. Under such circumstances, a duplicate (sister) tool would be called up, assuming that one has already been loaded into the tool magazine. A warning would then appear on the MCU's VDU indicating to the operator that the worn tool needs to be resharpened or replaced.

9.2.2.6   Automatic tool setting and in-cycle gauging

Measuring tool offsets and entering the appropriate values into the MCU can be a time-consuming task, particularly in a CNC machine tool fitted with 100 or more different tools. Ways of carrying out some of this work in a special tool presetting area away from the machine tool have been developed, but wherever the work is carried out it is still manually time-consuming. At the beginning of the 1980s a novel type of measuring device, called a touch trigger probe (TTP), was developed. This opened the door to automatic tool setting at the machine tool, as well as offering the ability to measure component features automatically by including canned measuring cycles at various stages in the part program.

The touch trigger probe is described in detail in Chapter 11 (§11.6.1.3), but for the purposes of this chapter it is sufficient to define it as a highly accurate sensing switch which, when actuated, transmits an electronic signal back to the MCU so that the controller can deduce its exact location with respect to a fixed datum point on the machine.

How this is achieved in practice is simplistically illustrated in Figure 9.14, but is described in more detail in current literature (Waters 1994). Suffice it to say that the time required to set just ten tools manually on a CNC lathe is typically 20 minutes, but by using a touch trigger setting probe this can be carried out automatically in approximately 2 minutes, and without risk of operator error.

By fitting a TTP into one of the machine's tool magazine locations, the probe can be automatically loaded into the machine's spindle in response to an appropriate instruction in the part program. It can then be used to measure the desired feature on the component and feed back this data to the MCU, which then automatically makes any changes to tool compensation or offset values necessary to produce an accurate finished part (Fig. 9.15). This is called in-cycle gauging.

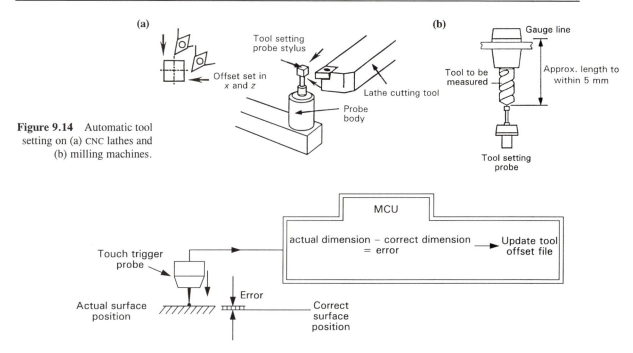

Figure 9.14 Automatic tool setting on (a) CNC lathes and (b) milling machines.

Figure 9.15 In-cycle gauging technique.

Exercise What vital instruction must the MCU issue to the machine tool's main spindle drive motor before a TTP is loaded in the machine's drive spindle?

Hint Consider the normal operating mode of the drive spindle.

Can you think of any other ways that a TTP could be used to further increase productivity of a CNC machine tool?

9.2.3 The machine tool itself

While CNC is now applied to a vast array of manufacturing machinery ranging from various types of machine tool to sheet metal, welding, pipe bending and precision measuring machines, by far the largest number fall into just two categories of machine tool – *machining centres* and *turning centres*.

A brief examination of these two types of machine tool would at first sight suggest that machining centres are effectively CNC milling machines and turning centres are CNC lathes (Fig. 9.16a and b). However, all is not what it seems, and CNC machine tools are anything but conventional machines with an MCU attached.

CNC machines differ from their conventional counterparts in a number of ways, the most important being that they are structurally much more rigidly designed. This is vital if consistently higher accuracy is to be achieved, particularly as modern CNC machines are normally fitted with much larger spindle drive motors than equivalent conventional machines. Coupled with the latest cutting tool technology, this enables greatly increased rates of metal removal to be achieved without sacrificing accuracy or repeatability.

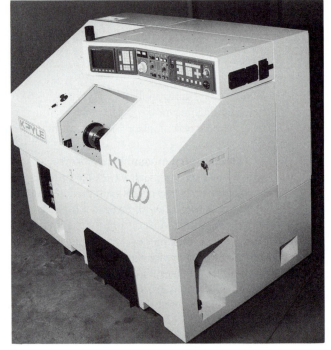

Figure 9.16 (a) CNC machining centre. (b) CNC turning centre. (Courtesy of Kryle Machine Tools, Stoke-on-Trent.)

Modified slideway designs are also used. These are much stiffer and offer enhanced vibration damping characteristics and are also usually angled to assist in swarf removal. Swarf can be a major problem with CNC machines because of their capacity to remove large volumes of metal per unit time. CNC slideways must also be designed in such a way that they have extremely low levels of sliding friction.

As the spindle and axis drive motors are all controlled by the MCU, it is essential that they be electronically completely compatibility with the MCU's control logic. For this reason most machine tool builders favour using motors provided by the MCU supplier.

Another vital element in the construction of CNC machine tools is their high-accuracy electronic measuring systems, but before explaining how these work the principle of closed-loop control – a basic positional operating system used by almost all CNC machine tools – is briefly described.

9.2.3.1 Closed-loop control

If an MCU sends an instruction to one of the machine tool's axis feed motors telling it to traverse to a specified position, it is essential that the drive motor keeps moving that axis until the desired position is reached, and then all movement must stop immediately. For this to be possible to the levels of precision and repeatability (down to 1 μm) demanded closed-loop feedback control of modern CNC machines is essential. While this sounds complicated, it is fundamentally a simple concept, and is used in many areas of industry where accurate control is required.

On machine tools, having issued a command to the appropriate axis control motor to traverse its axis to a given position, the MCU then continuously monitors the position of that axis and compares it with the final desired position. When the difference between these two values is zero, the required position has been reached and the MCU stops the drive motor (Fig. 9.17).

However, the success of this control process relies upon a succession of accurate axis positional values being fed back to the MCU, and this can only be achieved by fitting each axis of the machine with an extremely accurate linear measuring system.

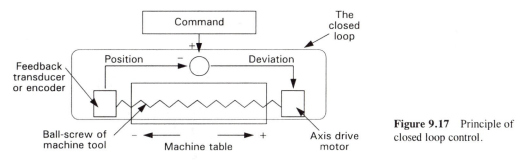

Figure 9.17 Principle of closed loop control.

9.2.3.2 Linear and radial measuring systems

Fitted to one end of each axis ball-screw (an ultra-low-friction, zero-backlash version of the lead-screw used in conventional machine tools) is the axis drive motor (Fig. 9.17), and at the other end is fitted a device called a rotary encoder (Fig. 9.18). This is the heart of the measuring system.

By counting the number of revolutions that it makes for a given ball-screw pitch, the encoder can determine how far the axis has travelled. It then sends this information electronically back to the MCU. The resolution of encoders is such that distance is computed for encoder angular movements as small as 0.0001 of one revolution (2 minutes of arc), and this equates to a linear incremental movement for a typical ball-screw of approximately 1 μm.

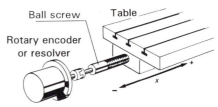

Figure 9.18   Typical location of a rotary encoder.

However, this only measures how far an axis travels, and is therefore synonymous with incremental programming (§9.2.1.2); accordingly it is called an incremental measuring system. Absolute measuring systems use similar encoders, but all measurements are taken with reference to a fixed zero reference point – usually set by the machine tool builder during commissioning tests – and is automatically re-established each time the machine is switched on.

Exercise   Many conventional machine tools are still sold, but most are now fitted with digital readout of axis movement (Fig. 9.19) rather than the age-old method of using a precision-engraved linear scale fixed to each axis. Why do you think digital readout is preferred, and does it increase the inherent accuracy and repeatability of the machine tool?

It is also often necessary to measure radial position accurately, a good example being screw cutting: if, after the first cut, the cutting tool does not enter the thread groove at exactly the same radial point at the beginning of each subsequent cut, no thread will be produced and expensive tooling will be broken.

Figure 9.19   Digital readout fitted to a conventional centre lathe. (Courtesy of Heidenhain (GB) Ltd, Burgess Hill.)

Digital readout unit

To measure precisely the radial position of a rotary axis (a common one being the rotary axis C about the z-axis on a turning centre), a device similar to a rotary encoder is attached to the rotating spindle; this is normally called a resolver. Its method of angular measurement is similar to a rotary encoder, but it is capable of angular incremental resolution as small as 10 seconds of arc.

Precise angular measurement is also essential when milling operations are carried out on a turning centre using power tooling (Fig. 9.20).

9.2.3.3 Power tooling

Turning centres, like any lathe, normally feed a stationary tool into the rotating workpiece. This is satisfactory for components that are circular about their rotational axis or if holes are required along the machine's rotational centre-line. However, if features that cannot be produced by such relative motion are required, then, after all turning operations are completed, it is necessary for the part to pass to other forms of machine tool, such as a milling machine or drill. With the advent of power tooling, these separate operations can often now be avoided.

A power tool, as its name implies, is tooling that has its own power supply, which enables it to rotate independently of the main rotational axis of the machine tool. With the component still clamped in the machine's chuck, but of course not rotating, the turning centre can then be programmed to bring in such power tooling to any required position relative to the workpiece. In this way machining radial holes, flats, etc. is possible, and even cam-type profiles can be produced by controlling precision feed in the C-axis relative to movement of the power tool in the x-axis (Fig. 9.20).

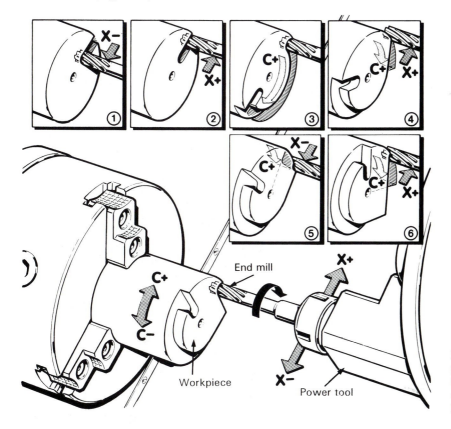

Figure 9.20 Use of power tooling on a turning centre. (Courtesy of H. Kief, Michelstadt, Germany.)

In many cases, by judicious planning, quite complex parts can be completed entirely on the turning centre. This avoids the time and cost of setting up a succession of secondary machines and the queuing delays that inevitably occur as batches of parts pass from one machine tool to another. Thus, productivity can be dramatically increased with a modest capital outlay in power tooling.

9.2.3.4 Automatic tool changers

For all but the simplest of parts, CNC machine tools use a wide range of cutting tools, which must be inserted into the machine's spindle in a preset sequence as specified by the part program. This can be carried out manually, but most modern machines are fitted with a tool magazine and automatic tool changer mechanism to allow the correct tool to be called up and inserted into the machine spindle automatically. On the latest machining centres a tool can be removed and a new one inserted in just over 1½ seconds, so maximizing the time that the machine is actually cutting metal.

Magazines in the form of an endless belt (Fig. 9.21) with up to 240 tool stations are available. They offer adequate capacity to cover all tooling needs for a number of different jobs, plus a number of sister tools for the most frequently used ones, but manually locating the correct tool at the instant required is not practicable with tool stores of any significant size and can only be managed efficiently by an MCU. As stated in section 9.2.2.5, the MCU can conveniently combine this control activity with automatic tool life monitoring.

9.2.3.5 Adaptive control

Very few machine tools are programmed to remove metal at anywhere near their design limits. Because the prime task of any machine tool is to remove metal, having due regard to accuracy constraints, it should be run as near as possible to its maximum capacity for as long as possible.

Unfortunately, metal is frequently not homogeneous, forgings and castings being particularly notorious for having tough skins and hard spots. If the programmer therefore calls for the maximum feed rate permissible for a given cutting tool, then immediately such a hard spot is encountered the tool will be overloaded, as could the machine's main spindle motor. To avoid this, programmers usually specify cutting parameters significantly below the maximum to allow for a comfortable margin of safety. However, the use of adaptive control (A/C) is a far more elegant solution to this problem.

A/C is a closed-loop control system capable of sensing cutting conditions at the machine's spindle by continuously monitoring spindle power and torque and spindle motor temperature. Instead of stating maximum permissible feed rates, the programmer specifies maximum acceptable tool loadings in terms of feed force, axial thrust force, spindle cutting power or torque. The A/C then, in conjunction with the MCU, continuously adjusts the feed rate so that the measured load parameters are kept as near as possible to their programmed values but never exceed them. Liberal safety margins are therefore no longer necessary.

A/C is also an efficient means of monitoring tool breakage.

Note   There is a practical limit to the level of sensitivity that can be built into adaptive controllers, and the larger their power rating the less is their absolute sensitivity. Can you suggest why this could be a significant problem when using A/C across the full range of machine tool spindle power?

Another area where A/C can be effective in optimizing the cutting process is in the machining of components whose shape involves air cutting (Fig. 9.22). Clearly, the machine's maxi-

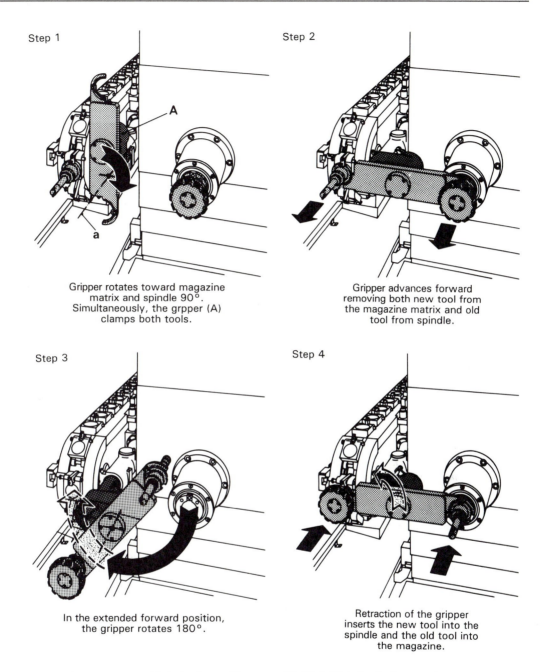

Step 1

Gripper rotates toward magazine
matrix and spindle 90°.
Simultaneously, the grpper (A)
clamps both tools.

Step 2

Gripper advances forward
removing both new tool from
the magazine matrix and old
tool from spindle.

Step 3

In the extended forward position,
the gripper rotates 180°.

Step 4

Retraction of the gripper
inserts the new tool into the
spindle and the old tool into
the magazine.

Figure 9.21 Automatic tool changer with duel gripper. (Courtesy of H. Kief, Michelstadt, Germany.)

mum feed rates can be used when traversing air gaps, but when the next metal surface is reached normal feed rates must be resumed immediately. Rapid response and sensitivity are essential if tool breakage is to be avoided, and this accounts for the relatively high cost of A/C systems.

261

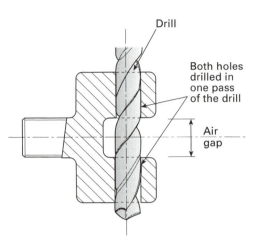

Figure 9.22 Example of a
part involving air cutting.

9.3 Flexible manufacturing cells and systems

The availability of reliable CNC machine tools at competitive prices, coupled with their great inherent flexibility, has resulted in increasingly widespread use worldwide. A logical next step was to link two or more together by a central control system to form a flexible manufacturing cell (FMC). This is usually referred to as cellular manufacture (see also §10.2.2).

There is no precise definition of an FMC, but they are generally considered to comprise one or two stand-alone CNC machines complemented by ancillary equipment, such as robotic parts handlers, to allow unmanned operation for a limited period (Kief & Waters 1992). They are usually only capable of machining a restricted family of components.

For FMCs to function efficiently the following are necessary:
- an adequate supply of parts to last a complete shift – usually 8 hours
- automatic loading and unloading of parts to and/or between the CNC machines
- automatic cutting tool monitoring of tool wear or fracture, with the ability to call up replacement sister tooling as required
- the ability to monitor and control the dimensions of the machined part by touch trigger probing, with automatic feedback to correct for minor deviations or ultimately to stop the cell in the case of unacceptable deviations
- the ability to stop the cell automatically when the entire supply of parts has been machined.

Flexible machining systems (FMS) are an extension of the FMC principle and usually include rather more CNC machines together with such additional features as a common materials handling transport system, and possibly a co-ordinate measuring machine (CMM) (see §11.6), frequently preceded by an automated component wash/dry unit (Bonetto 1988). The whole system is controlled from a central computer, which exercises overall system control, as well as providing a DNC centralized part program storage facility.

An FMS can be likened to a flexible transfer line in that it offers high productivity but at the expense of the great job-to job-flexibility offered by stand-alone CNC machines (Fig. 9.23). This is discussed in more detail in Chapter 10 (§10.2.3).

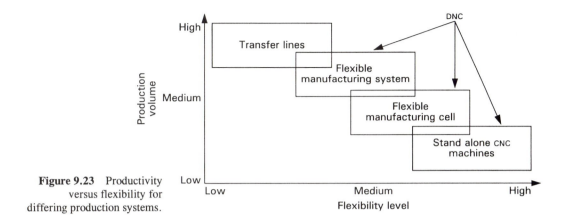

Figure 9.23 Productivity versus flexibility for differing production systems.

Nevertheless, within the constraints of the range of parts for which an FMS has been programmed, it can cope cost-effectively with single parts machined in random order, i.e. no minimum batch size is necessary. It also lends itself to adaptation to different workpiece family configurations without the lengthy set-ups inherent in pure transfer lines (§10.2.3), although frequently there is a need for significant changes to the system's transportation layout. However, this is not usually a major problem as the most popular forms of transport system comprise either rail- or inductive loop-guided self-propelled carts (Fig. 9.24) called automatic guided vehicles (AGVs). Rerouting the rail or inductive loop path, while temporarily disruptive to production, is technically not difficult.

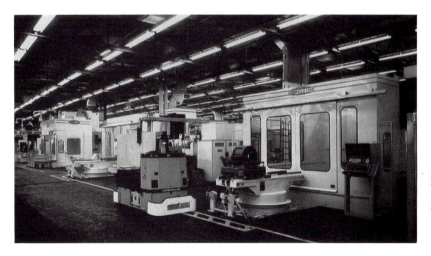

Figure 9.24 Modern FMS installation showing an AGV controlled by an underfloor inductive loop. (Courtesy of Giddings & Lewis, Prescot.)

9.4 Summary of the principal benefits of CNC machine tools

(a) Greatly improved accuracy and repeatability is achieved compared with conventional machine tools.

(b) Once a part program has been used satisfactorily it can be re-used subsequently with guaranteed success.

(c) While CNC machine tools can be manually programmed via their MCU, programming can be carried out away from the distracting shop floor environment and, depending upon component complexity, one part programmer can usually satisfy the programming needs of a number of CNC machines.

(d) Consistent production times are achieved as human fatigue is not a factor.

(e) Set-up times are minimal, making batch sizes of even one part economical.

(f) Depending upon component machining times and batch sizes, more than one CNC machine can be supervised by a single skilled operator.

(g) It is possible to link CNC machines with automatic workpiece handling, making the periods of unmanned operation even longer.

(h) Groups of CNC machines can be readily controlled by a single host computer to form a fully integrated system.

(i) Well programmed CNC machine tools spend 70 per cent or more of the working day actually cutting metal; manually operated conventional machine tools are lucky to achieve a rate of 10 per cent.

(j) The use of automatic tool changers and touch trigger probing for tool setting and in-process gauging dramatically reduces non-cutting time.

(k) Automatic in-process inspection can be included in the part program, greatly reducing the likelihood of scrap components being produced.

(l) Optimum cutting conditions can be maintained as the machine tool's operating parameters are controlled by the part program rather than by an operator. The use of adaptive control further increases the efficiency of the cutting process.

(m) With power tooling, complex parts can be made in one setting of a machine tool, thus avoiding the need to pass batches of part-finished components from one machine tool to another before they are complete.

(n) Production planning and control is simpler and more reliable than is possible in manually operated machine tool workshops.

(o) Extremely complex shapes that could not be made in any other way are now possible, thanks to the ability of modern MCUs to control all the machine's axes simultaneously.

(p) Component fixturing is greatly simplified, and in many cases can be eliminated altogether.

(q) Component design changes can usually be accommodated by minor alterations to the part program. With fixtured manufacture typical of conventional machine tool production, costly modifications, or even new fixtures, may be required.

(r) If subcontracting becomes necessary owing to an unexpected shortage of machining capacity, transferring part programs is easy and ensures that the parts continue to be made in the same manner as in one's own factory.

(s) Because floor-to-floor machining times can be accurately predicted to within ±5 per cent at the programming stage, more accurate tendering is possible. It is also essential when operating just-in-time (JIT) manufacture (§1.2.2.2).

(t) New parts that have a number of similar features to previously programmed components can often be programmed quickly by making minor changes to the original program.

Bibliography

Bonetto, R. 1988. *Flexible manufacturing systems in practice*. London: North Oxford Academic Press.

Gibbs, D. 1987. *CNC part programming: a practical guide*. London: Cassell.

Kief, H. B. & Waters, T. F. 1992. *Computer numerical control*. Westerville, OH: Macmillan/McGraw-Hill.

Thyer, G. E. 1991. *Computer numerical control of machine tools*. Oxford: Newnes.

Waters, T. F. 1994. Productivity enhancement of CNC machine tools by the use of touch trigger probing. Paper presented at the International Symposium on Mechatronics, Mexico City, January.

Case study

A small manufacturing company decides to purchase its first CNC machining centre at a cost of £70 000. As their production manager you are aware that the great improvement in productivity that this new machine will bring could be significantly enhanced by an additional investment of £5000 to equip the machine with both touch trigger probe tool setting and in-cycle gauging facilities (§9.2.2.6). Unfortunately, the board of directors is reluctant to spend this additional capital. How would you persuade them that the additional investment is commercially worthwhile?

Solution

One of the first questions that should be addressed is "Why not wait until the new machine is operational and then use a little of the profit it generates to install touch trigger probing (TTP) at a later date?". While this is technically feasible, and is the path followed by many CNC users, it is always better to include as many features as possible into the initial operator learning period and while you have (free) technical support during commissioning and possibly for a finite period thereafter.

Furthermore, the machine tool builder is contractually obliged to ensure that the machine tool operates correctly as a complete package if TTP is fitted at the time of purchase. If TTP is fitted later (*retrofitting*) and there are problems in getting it to function correctly, then the probe installer could blame the machine tool and/or MCU and the machine tool builder could blame the TTP equipment or installer. Either way you are stuck with equipment that does not function properly – a situation to be avoided at all costs.

Let us next turn to the benefits of a tool setting probe. In the majority of CNC machine tools each of the cutting tools in the tool magazine is initially set up manually using such equipment as feeler gauges, dial test indicators, gauge blocks, and so on. This is both time-consuming and prone to operator error.

Fortunately, this is one of the easiest and economically most attractive operations to automate using TTP. Time savings can be enormous – up to 90 per cent – and when power tool (§9.2.3.3) setting is involved time savings are even greater because of the more complex calculations and setting procedures required.

To summarize, TTP tool setting facilities eliminate the production of scrap resulting from human set-up errors, automatically register each tool's offsets, abolish the need for trial cuts and greatly reduce set-up times, thus dramatically improving productivity. The tool setting probe can also be used to check automatically for broken tools by checking tool length before returning each cutting tool to the tool magazine.

With such potential gains in productivity, it is difficult to envisage any situation in which automatic tool setting could not be economically justified.

Finally, we will consider the economic benefits of in-cycle TTP gauging. In-cycle gauging involves holding a TTP in a standard tool holder and placing it in one of the machine's many tool holder stations. Thus, when the part program calls up the tool placed at the station containing the probe, the probe is then automatically loaded into the machine's spindle, ready for use.

Figure 9.15 illustrates the principle of in-cycle gauging, and section 9.2.2.6 provides a brief description, so no further comment is required here. However, there are a number of other highly productive functions that can be performed by a TTP located in the machine tool's spindle. These are:

(a) *Part identification*. Probing can be programmed to check key features on each newly loaded item to confirm both that it is the correct component and that it has been loaded correctly.

(b) *Surface verification of castings and forgings*. This helps to eliminate fresh air cutting (Fig. 9.22) and detects the possible presence of excess material.

(c) *Feature shift on cast components (e.g. cored holes)*. Probing casting features as they really are rather than as they should be enables stock removal to be as uniform as possible on every casting.

(d) *Automatic tool wear compensation*. This is synonymous with in-cycle gauging mentioned earlier.

Questions

9.1 How do CNC systems differ from earlier NC systems?

9.2 What are the three basic constituent parts of any CNC machine tool system?

9.3 What two types of data are found in any CNC program?

9.4 What are processors and post-processors as used in the operation of CNC machine tools and what are their principal functions?

9.5 What are G and M codes, and why does their use simplify CNC part programming both in the planning office and when manually programming at the machine tool?

9.6 What is the difference between modal and non-modal programming instructions and why is it essential to be aware of their difference when programming?

9.7 List as many functions as possible that are performed by the modern multiaxis MCU.

9.8 Why are straight-line and two-axis contouring methods of cutter control no longer of major significance?

9.9 What is cutter path offset and how does it differ from tool compensation?

9.10 What essential geometrical relationship must be maintained between cutter path and workpiece contour if the required surface is to be accurately produced?

9.11 The use of touch trigger probes has become a standard feature on many CNC machine tools. What are their prime functions in this capacity?

9.12 What are the main structural differences between conventional machine tools and their CNC equivalents?

9.13 What benefits does the use of power tooling offer the production planning engineer?

9.14 What are the objectives of adaptive control as applied to CNC machine tools? What is the greatest single limitation of A/C systems currently available?

9.15 How do FMCs and FMSs differ from one another and from stand-alone CNC machines? Why is an FMS frequently not the best solution for increasing productivity?

9.16 List what you consider to be the most important benefits offered by CNC machine tools if your main concern is: (a) increased flexibility, (b) increased productivity or (c) improved dimensional accuracy.

9.17 The whole structure of manufacturing industry is undergoing fundamental change. The conventional concept of efficient production is to manufacture by mass production, but more and more areas of industry are being forced to develop a radically new approach as increasingly the requirement of manufacturing is to produce varying quantities of an ever-widening range of products at the lowest possible unit cost. Discuss the fundamental causes and implications of this situation on manufacturing in the UK as we move into the next millennium. How has the development of CNC machines contributed to this change in manufacturing philosophy?

9.18 The plate shown in Figure 9.25 contains nine 5 mm diameter holes, the central five being reamed. Write a program using absolute co-ordinates to perform the hole drilling and reaming required, explaining the meaning of each program step. Include all tool changes and return the spindle to its initial park/tool change position. Rewrite the lines of your program concerned with the drilling of the four unreamed holes (A, B, H and J) but this time in incremental mode, again returning the spindle to its park position. Ensure that you observe the unusual axes convention shown.

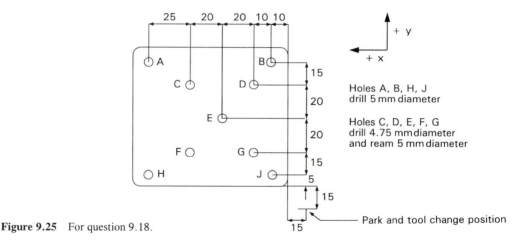

Figure 9.25 For question 9.18.

267

Chapter Ten
Productivity and automation

After reading this chapter you should understand:
- (a) what the four principal forms of manufacturing are and how the production planning and control requirements differ between them
- (b) why the way that equipment is configured on the shop floor influences manufacturing efficiency
- (c) the need to optimize equipment, material utilization and human resources if the most cost-effective manufacture is to be achieved
- (d) ways of monitoring manufacturing productivity levels
- (e) the difference between mechanization and automation
- (f) when and when not to use automated processes
- (g) areas to consider when contemplating automation as a means of increasing productivity
- (h) what an industrial robot is and what they are principally used for in a manufacturing environment.

Readers who have methodically worked through the previous chapters of this book should, by now, have a basic understanding of the principal processes used in manufacturing industry. While this is essential knowledge, knowing how to employ these processes in the most efficient manner is also vitally important if the business is to be both profitable and able to resist competition.

But what does efficiency mean in this context? Since an essential objective of any commercial enterprise is to make a healthy profit, in the manufacturing context efficiency implies making a product to the required specification, on time, and at the lowest possible cost. To achieve these goals the company must therefore make optimum use of its equipment and human resources, as well as ensure that raw materials are used to best advantage. This is not always as straightforward as one might imagine, because using the most suitable process available may not be the most efficient approach, as it may be possible to redesign the part, subassembly etc. in such a manner that production costs can be lowered by a greater extent than would be possible by simply optimizing the process.

Indeed, one of the greatest changes in manufacturing in the past 20 years has been the increased awareness of industrial designers of the costs of manufacture. "Design for Manufacture" has now become an essential part of most engineers' training, and involves topics such as value engineering and value analysis (Edwards & Endean 1990), whose purpose is to optimize design to achieve minimum manufacturing costs without jeopardizing product quality or functionality.

Note "Value engineering" and "value analysis" have broadly the same objectives, but value engineering is the term used when referring to the design of new items and value analysis normally relates to the redesign of existing products.

Another aspect that must be considered when planning for optimum manufacturing efficiency (productivity) is the immediate and likely longer term quantity requirements, as this can have a significant impact upon future capital expenditure.

To appreciate how production volume requirements influence equipment layout, it is necessary to gain an overview of the four manufacturing formats most commonly used by industry.

10.1 The principal manufacturing formats

The following four types of manufacture are very broad in that there is no precise quantity break-even point between one category and the next. However, the numerical values shown in Figure 10.1 are an approximate indication of typical values involved.

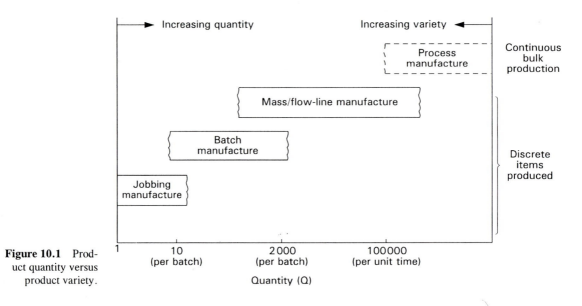

Figure 10.1 Product quantity versus product variety.

The other clear division of manufacturing formats is between process production, whose output is continuous (§10.1.4), and all other forms of manufacture, whose output involves the production of discrete items.

10.1.1 Jobbing production

Although this covers the one-off situation, the product can range in size and complexity from a ship or a road bridge to small items of hand-made jewellery. The job is usually tendered for and, if successful, is made to order. This requires a dynamic technical sales organization to achieve rapid, realistic tendering.

270

A wide range of general-purpose machines and equipment is needed to maximize the number of jobs that can be tendered for. Highly skilled personnel are also required, both on the shop floor and in the design office, to cope with the wide spectrum of jobs that are likely to be encountered and the inevitable changes/modifications typical of this type of manufacture. Thus, precise production control is more important than highly detailed production planning.

Machines are usually laid out on a function basis (§10.2.1) for maximum flexibility and large component and raw material storage capacity is normally required because of the inevitable high level of work-in-progress.

10.1.2 Batch production

This is the manufacture of similar articles made in batches to meet a continuing sales demand. Batch sizes vary widely, but will usually be in excess of immediate sales requirements, the balance being stored until required. An efficient stores management system is therefore essential.

Machine tools are of the general-purpose type, usually arranged functionally (§10.2.1), but a limited amount of specialist equipment is also frequently incorporated. Cellular layouts (§10.2.2) can sometimes be used to great effect.

Much work-in-progress is always involved, although this is reduced in cellular layouts, as is the movement of work within the factory.

Production planning and scheduling (Buffa & Sarin 1987) must be as efficient as possible to minimize the number of machine set-ups required, and up-to-the-minute production control is also essential if late delivery problems are to be avoided.

Costing is likely to be more accurate than with jobbing manufacture as historical data from previous batches is usually available.

10.1.3 Mass production

When very large volumes of identical items are required on a continuous basis, e.g. domestic white goods and cars, equipment is laid out to provide a balanced flow of work from one process to the next, with materials being fed into the system at a preplanned rate as near as possible equal to sales demand. Work-in-progress is thus at an absolute minimum. Work movement is also minimized with this type of production, and what there is usually employs a high level of mechanization.

Most manual labour tasks are deskilled by the use of sophisticated machinery, although such highly specialized equipment and machines have poor flexibility and to alter the product line in any way is very expensive. However, this is changing thanks to the integration of reprogrammable robotic devices (§10.9) into the production line.

Failure of any part of the system can have disastrous effects on production, although buffer stocks (§10.2.3) can be used to help alleviate this.

Production planning must be absolutely correct as subsequent changes are both expensive and difficult to implement once mass production has started.

Production control requirements are minimal as most operations are performed in an automatically controlled manner. However, rigid planning and control of incoming and outgoing goods is essential to avoid the need for massive storage facilities.

Exercise   A large car manufacturer produces a normal four-door saloon car every 2 minutes. Make a list of the principal bought-in components required (e.g. tyres, electrical components, etc.), and then calculate how many of each are needed per week. Assume that the plant is operating at full production 24 hours per day, 6 days per week. Now estimate the size of the storage areas required to contain just 2 days' requirements.

Costing is usually very precise because of the on-going repetitive nature of this type of production.

10.1.4 Continuous process production

This involves the manufacture of a continuous stream of product and, with the exception of the continuous casting of steel (§2.5.5) and certain forms of plastic manufacture, is rarely encountered in mechanical engineering. It is more the province of the chemical and food processing industries, in which such products as fuel oil and cereals are produced by continuous flow.

10.2 Equipment layout

The relative positions of manufacturing plant such as machine tools can have a significant effect on productivity. There are four broad categories of layout, the most suitable one for any given situation being greatly influenced by the type of production involved (§10.1).

10.2.1 Functional layout

This is one of the oldest and most common layouts and involves placing all machines of a given type together in one area (Fig. 10.2).

Figure 10.2 Functional layout.

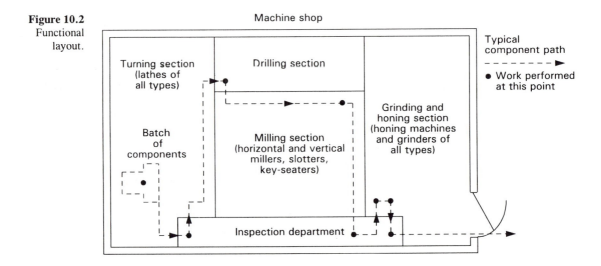

Specialized supervision can be readily provided in this system by charge-hands and foremen who have specialist knowledge of the type of machines under their control and of the jobs that pass through their area.

Flexibility is good, and this layout is ideal for both jobbing and small batch manufacture. However, there is usually a large volume of work-in-progress, and throughput times can be high because of the significant mechanical handling of parts in moving them from machine to machine and from one group of machine tools to another (Fig. 10.2).

10.2.2 Group/cellular layout

This format, while nothing like as common as the functional layout, has become increasingly popular with the introduction of CNC machines. It involves placing together in a cell a group of machines that are needed to make a specific component or family of components (Fig. 10.3).

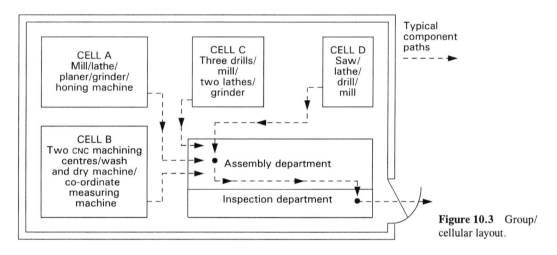

Figure 10.3 Group/cellular layout.

Providing the volume of production justifies it, this layout offers a number of distinct advantages, such as:

- The group supervisor can reduce administrative paperwork to a minimum, and will ensure that the product leaving the cell is to the desired quality, i.e. delays and errors are not blamed on people within the cell.
- Cellular working fosters team spirit, i.e. all members of the group are likely to pull together to ensure that it is not their group that causes delays or quality problems in subsequent operations.
- Within the group there is much less likelihood of the "its not my job" attitude.
- As members of a group are usually capable of performing all the tasks within the cell, this makes job rotation possible and so reduces the risk of boredom.

While there may be a fair degree of flexibility within a particular cell, this is often not the case between cells, i.e. machines in any given cell are there to produce one particular product or family of parts only, and any radical product change may result in the need for totally new machine groupings. Also, maximum machine utilization is not guaranteed. For example on a particular day, the grinding machine in cell A (Fig. 10.3) might be only partially employed while the grinder in cell C may be temporarily overloaded owing to a previous machine breakdown.

10.2.3 Flow-line layout

When a continuous demand for the same or similar product is assured over an extended period (as is typical of the automotive industry), it may be economically worthwhile installing a specially designed system of machinery, called a *transfer line*, to perform a fixed number of tasks. Automated mechanical handling between operations is usual: the raw material, in the form of bar, casting, forging, etc., enters the system at one end and the finished part appears at the other end with little or no human intervention.

This is known as flow-line production, and while many such systems are in a straight line (Fig. 10.4a) it is also possible to have a circular or U-shaped layout (Fig. 10.4b) with a consequent reduction in the floor area required.

Frequently a number of lines are working concurrently and their outputs fed to an end product assembly line area.

Failure of just one operation would rapidly cause stoppage of the whole line; to minimize this risk, a small stock of parts – termed a buffer stock – is held at what are judged to be points of particular vulnerability (Fig. 10.4a and b).

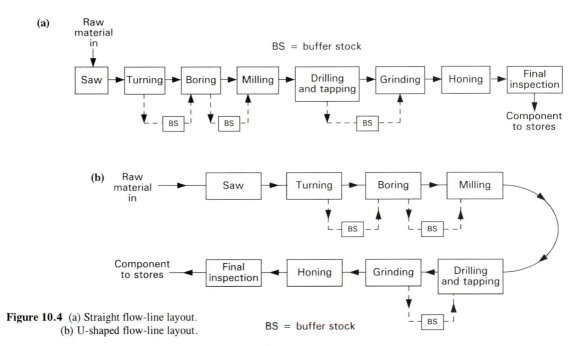

Figure 10.4 (a) Straight flow-line layout.
(b) U-shaped flow-line layout.

Flow-lines offer the greatest productivity levels of any system yet devised, but its inflexibility and major capital cost make it a high-risk strategy. This is particularly true as consumer demand moves more and more towards increased customer choice and shorter product life cycles (§1.2.2).

Another requirement of flow-lines is that all their individual operations must be balanced. This implies that each operation in a flow-line should take the same time to complete. In reality, different operations have different cycle times; as it is pointless to be always waiting for the slowest operation to be completed, it is usual either to adjust the speeds of all the other operations so that they equal the speed of the slowest operation or effectively to halve the time required by the slowest operation by duplicating it on the production line.

In an attempt to introduce a degree of flexibility into flow-lines without significant loss of productivity, reprogrammable robotic devices and CNC machinery are increasingly being introduced into such systems. This also greatly shortens the time (usually measured in months) required to modify the line to make it capable of producing a new product.

Exercise   When a car factory is working at full capacity it may produce a car every 2 or 3 minutes. Clearly, it is not possible to build either a complete engine or a gearbox in such a short period of time. How do you think that this problem is overcome?

10.2.4 Fixed product location layout

In the previous three layouts the product or component moves past fixed-position production machinery. Here the reverse applies, and in some cases neither the partly completed nor even the finished product ever moves. This is typical of many civil engineering projects, such as bridge construction. Another example is in shipbuilding: ships are not moved until at least their basic structure is completed (Fig. 10.5).

Figure 10.5
Assembly of a ship's hull. (Courtesy of Vosper Thorneycroft (UK) Ltd, Southampton.)

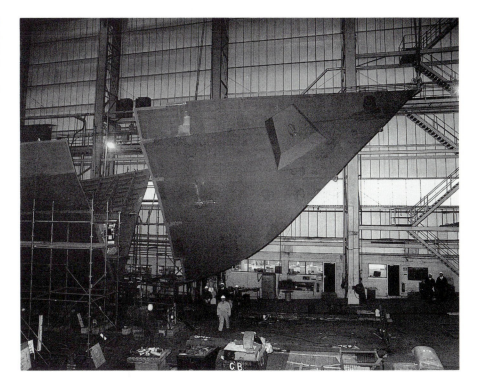

10.3 Optimizing material use

When faced with the task of making a component in a given material, depending upon such factors as quantity, physical size and shape, the most suitable procedure is usually immediately obvious to an experienced planner. What may not be so clear-cut is the possibility of minimizing material use.

As swarf (scrap metal) is at most worth only half of its initial value, there is a large financial incentive to optimize material utilization as a contribution to minimizing overall manufacturing costs. [This is why "nesting" of sheet metal parts (Fig. 3.27) is so worthwhile.]

The planner should therefore take the view that any material used in excess of the mass of the finished part is lost potential profit, and with this in mind examine whether modification of component design (value analysis) or judicious planning to optimize material use will minimize material requirements.

As is frequently the case in engineering, compromise may be called for, because a change in the design or manufacturing technique might well reduce the amount of metal used but at the same time increase the costs of the manufacturing processes involved. An example of such a situation is the subject of the case study at the end of this chapter.

Clearly, each situation must be examined on its merits, but any steps that result in improved material utilization, whether design modifications or imaginative production planning, are worthy of careful consideration.

10.4 Use of human resources

Effective employment of people to assist in the goal of maximizing productivity is possibly the most difficult aspect of production management. Both personal and national characteristics are a significant factor, as can be seen when comparing, for example, the attitudes to work and their company of British and Japanese workers.

Successful managers learn to motivate staff for the benefit of both company and employees by using a variety of incentives. The most obvious of these is financial, although a pride in one's work, a feeling of being a valued member of a team and the provision of a good working environment all play an important part in maintaining a contented workforce.

However, as long as people are involved in the actual manufacturing process, without elaborate automatic checking systems in place, human errors will always occur. The only way to eliminate this risk is to remove direct human involvement and replace it with automated processes (§10.6). This approach also has the added advantages of eliminating variations in worker productivity levels and cost differences between high and low labour rate economies.

Exercise   Make a list of other non-financial incentives that are likely to enhance worker productivity. Why do you think that significant variations in productivity will always exist between employees, even among those who are equally conscientious?

10.5 Measuring manufacturing productivity

The most obvious yardstick by which productivity can be measured is by the ratio of output to works cost. However, this is far too simplistic and calls for rather more detailed examination.

Finance in its broadest sense must always be the ultimate basis upon which any manufacturing enterprise is assessed, because if its costs are not less than its income it will not survive. But what is meant by cost?

Every company has its own method of assessing incurred costs, but in the end all direct and indirect costs involved in running a business have to be reflected in the selling price of its products. Because there is a limit to the price that the customer will be willing to pay, because of competition and/or perceived value, it is usually not possible to apportion overheads (indirect costs) uniformly across all of the company's products. Examination of the many options open to cost accountants in apportioning overall costs is outside the scope of this book, and it is only necessary here for the reader to appreciate that *all* costs, however incurred, must ultimately be paid for by the customer.

Exercise   Most simply, direct costs can be considered to be costs that are directly attributable to the production of a specific product. Indirect costs (or overheads as they are often called) are all the other costs incurred in running a business. List as many activities and costs as you can that would normally be included in the overheads of a medium-sized manufacturing company.

As manufacturing engineers it should be our constant aim to maximize added value at every stage of production. A frequently quoted parameter (Wild 1989) based upon this philosophy is:

$$\text{Output per employee} = \frac{\text{added value per unit time}}{\text{total number of employees}}$$

Added value here includes all company direct and indirect costs but, as indirect costs are frequently as much as 300 per cent of direct manufacturing costs, when quantifying their productivity performance production engineers are better advised to use other parameters more directly related to their spheres of influence. Good examples are utilization of equipment and materials, reject costs per unit time, ratio of late deliveries to output per unit time, value of work-in-progress to value of output per unit time, etc.

10.6 Mechanization and automation

Non-engineers invariably make no distinction between the terms mechanization and automation, which is hardly surprising in view of the wide range of definitions quoted in dictionaries and technical literature. Nevertheless, they are different and, while the following are not claimed to be *the* definitions, they will hopefully provide the reader with an appreciation of their basic difference.

Mechanization of an activity is the replacement of human effort by a mechanism. Good examples of this are the automatic feed on a conventional machine tool and cams used to open and close valves at the required time and velocities in an internal combustion engine.

Automation also involves the replacement of human effort by mechanical devices, but it also incorporates, to a greater or lesser extent, an element of process control. It is therefore difficult to imagine any automated process that does not include mechanized elements.

Examples of automation are CNC machine tools fitted with in-cycle gauging (§9.2.2.6) or adaptive control (§9.2.3.5), and FMSs (§9.3). These examples have one characteristic common to all truly automated processes, namely they are all closed-loop systems (Fig. 9.17) in which feedback provides the necessary data for effective automatic process control. It can therefore be said that truly automated systems transform the operator's role into one of supervision.

Note   If you cannot recall what is meant by a closed-loop system reread §9.2.3.1.

Automated systems can incorporate many interconnected processes, such as automatic materials handling within and between workstations, automatic inspection, automated assembly, automatic testing and even automated packaging. Because modern automated systems are computer controlled, this makes the provision of on-line process data to system supervisors a simple matter.

Increasingly, computerized automated systems have a inbuilt diagnostic capability that is frequently able either to self-rectify faults or, alternatively, to call for assistance, giving the supervisor a precise indication of the fault.

Finally, before leaving the explanation of automation, the reader should note that there are two basic forms of automated system: fixed and programmable.

10.6.1 Fixed automation

When a system has been designed specifically for one purpose and would require major modification before it could perform some other task, it is termed a fixed automation system. Classic examples of such systems are special-purpose automatic machines and the flow-line type of mass production used in the automotive and domestic white goods industries (§10.2.3). Such systems are also referred to as hard or rigid automation, because many of the mechanical movements involve rigid devices such as cams, levers, end stops, etc.

10.6.2 Programmable automation

As its name implies, programmable systems are largely computer controlled and can therefore be relatively easily reprogrammed to modify their activities. Investment is usually much less than with special-purpose fixed automation machinery, as standard pieces of equipment can normally be brought together to form the required system; frequently the only problem is getting the computer controls of each element of the system to communicate fully with one another.

Because of the inherent flexibility of such systems (FMSs being a good example; see §9.3), small batches are economical (Fig. 10.6).

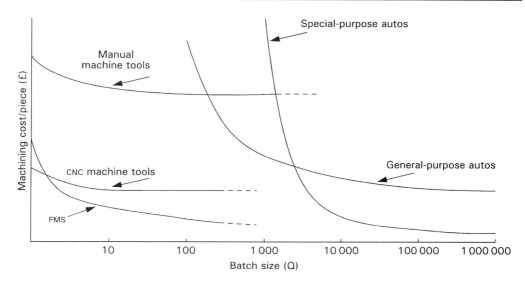

Figure 10.6   Machining cost versus batch size.

10.7 Reasons for and against automation

In addition to increasing output radically, when any significant level of automation is introduced into a manufacturing plant it is also likely to have a major impact on the workforce directly involved. It is therefore worthwhile considering some of the more important aspects associated with the implementation of an automation policy.

Its first and most noticeable effect should be to increase output dramatically, particularly as 24 hour, 7 day a week operation becomes possible with minimal staff being employed at antisocial hours. Indeed, it is often essential to operate the equipment continuously to justify economically the high capital costs involved.

Labour costs are dramatically cut, although a technically more skilled workforce is normally required. This can be of considerable assistance if there is a shortage of skilled labour available. While automation can lead to redundancy, it rarely does so in the engineering industry, and over-manning is usually overcome by both natural wastage and retraining to a higher skill level.

Exercise   Can you see numerous similarities between the implications of introducing automation and CNC machine tools onto the shop floor?

With reduced staffing levels, and with their largely supervisory role, workers are likely to become more socially isolated at work. This is further aggravated by the increase in shift working that is normally required.

Understandably, people are reluctant to work in certain types of environment. Indeed, some conditions can be positively injurious to health, and automation may then be the only option.

By drastically reducing human error, conformance to specification (§11.4.1), and consistency of product quality, greatly improves. This, in turn, results in less scrap and minimizes the risk of late delivery.

With the trend towards customer-initiated manufacture (e.g. JIT; see §1.2.2.2), the ability to predict precise manufacturing times is essential. Furthermore, with flexible programmable automation the time from design to completion ("lead time") is dramatically reduced. This also results in financial savings resulting from reduced levels of work-in-progress.

All processes are guaranteed to be carried out at the programmed optimum conditions and cannot be changed without proper authority. Thus, production management has real-time control over the manufacturing process at all times.

Certain computer-designed parts are so complex that they can only be made on equally complex computer-controlled equipment. Modern gas turbine blade and aircraft wing profiles are good examples.

On balance, the reader can be forgiven for concluding that a sound strategy would be to automate as many areas as technically possible. However, this is not at all the case and directly contravenes the most important automation strategy rule, which is: *only automate processes if it makes economic sense to do so*. This may seem obvious, but there are many examples of automation being introduced for automation's sake when it would have been economically much more sensible to have continued to use manual labour.

10.8 Some useful automation strategies

From §10.7 the reader should have gained a basic understanding of when the introduction of some level of automation is justified and when it would not be worthwhile. However, because the capital costs of implementing automation are considerable, it is essential to carry out a detailed analysis of both current and proposed manufacturing methods to ensure that the most cost-effective system is adopted. It is also important to ensure that, without jeopardizing either performance or quality, the current design is simplified as much as possible (Mather 1989). This is part of the value analysis exercise referred to earlier (§10.3).

While the following are examples of typical questions asked by planning engineers when carrying out such manufacturing cost optimization exercises, there is no substitute for designing products specifically for automatic manufacture/assembly (Stanton 1990).

- Can a number of operations be performed at the same workstation simultaneously? An example of this is replacing a series of drilling operations with a single multispindle drilling unit.
- Can a number of workstations be combined into one integrated system by the use of automatic mechanical handling devices? This allows more than one component to be worked on simultaneously and is common in flow-line manufacture (§10.2.3).
- Is it worth designing special-purpose machines to perform one or more operations with the greatest possible efficiency? This is an extremely expensive option and product quantities and characteristics have to justify such a high-cost approach.
- Can flexibility be incorporated into equipment design without jeopardizing productivity levels? This is usually achieved by the inclusion of reprogrammable CNC elements, and is an increasingly popular strategy in the automotive industry.
- Can materials handling be improved by the introduction of automated robotic handling devices (§10.9), reducing the physical distances that parts have to be moved or designing them so that they are easier to pick up and manipulate?
- Can the process be performed more efficiently by use of such features as adaptive control (§9.2.3.5)?

– Is it possible to reduce set-up times? Time spent setting up machines is totally non-productive, and every effort should therefore be made to minimize it. Typical ways of achieving this are by using universal fixturing with common machine location points, introducing FMSs (§9.3) and improving job scheduling efficiency so that modifications to the set-up are minimized when changing over from one batch of components to the next.

– Is shop floor data collection as up to date as possible, so enabling manufacturing management to react more rapidly to changing situations?

– Is it possible to introduce a common manufacturing database for use by all persons involved in the manufacturing chain? This would include process planning, production planning and control, jig and tool design, material selection, etc., and is often called computer-aided manufacture (CAM) (§1.2.2.3). In more sophisticated systems computer-aided design (CAD) also inputs into this database so that computer-aided engineering (CAE) $\rightleftharpoons$ CAD + CAM.

10.9 Industrial robots

Whenever possible, relieving the monotony of repetitive human activity is usually worthwhile, if only to minimize the risk of mistakes and to maintain constant cycle times. When cycle time requirements exceed human ability or environmental conditions are unacceptable, there may be no alternative but to use some form of industrial robotic device (IR). Typical IR applications in manufacturing industry are shown in Table 10.1.

Table 10.1 Current robotic device applications in engineering.

Application	Percentage	Type of control
Welding	28	Continuous path
Surface coating such as painting and plasma arc spraying	21	Teach-in continuous path (§10.9.2)
Loading/unloading injection moulding machines	12	Pick-and-place
Loading/unloading sheet metal punches	19	Pick-and-place
Workpiece handling on machine tools	10	Pick-and-place
Miscellaneous	10	Various

IRs and CNC machine tools have a number of similar characteristics, and use the same numerical control technology. Indeed, their application and operating modes are often planned by the same production staff. The biggest single advantage that makes them so cost-effective is their inherent flexibility. A robot cannot, of course, replace a CNC machine tool, and usually only acts in a supportive role in order to increase the level of automation possible.

Normally IRs are freely programmable mechanical handling devices with several axes of freedom, and they are usually equipped with grippers or special-purpose tooling. Most IRs can be classified as either "pick-and-place" or "continuous path" units, depending upon how they function.

10.9.1 Pick-and-place robots

IRs that are only required to move from one end position to another are ideally suited to loading and unloading machine tools (Fig. 10.7) and for simple assembly operations in mass production.

They traverse in two, three or occasionally four axes, but the control of intermediate positions between programmed end points is not normally possible. Because only a simple numerical control is necessary they are the least expensive form of IR.

Axis movement is usually achieved using either pneumatic or hydraulic cylinders.

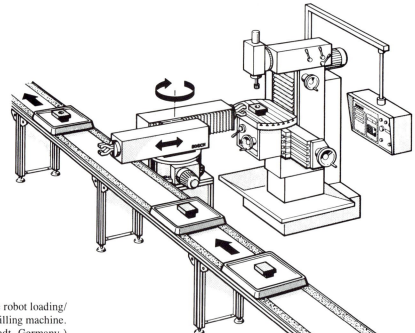

Figure 10.7 Pick-and-place robot loading/unloading workpieces on a milling machine. (Courtesy of H. Kief, Michelstadt, Germany.)

10.9.2 Continuous path robots

IRs that move along a defined three-dimensional path and are continually monitored by a closed-loop control system are called continuous path or universal robots. Because their movement in all axes can be continuously and independently controlled they are capable of mimicking many complex human movements and can perform such functions as fusion welding (Fig. 5.9) and paint spraying.

Writing the programs required to make IRs perform complicated multiaxis movements would be extremely difficult, so the usual way of programming continuous path IRs is by the *teach-in* method. This involves a human operator manually carrying out the required operation(s) while holding the IR's gripper (its hand). The movements are simultaneously recorded and then digitized. The program thus created can then be replayed at will, making the IR's hand reproduce exactly the original movements of the human operator.

Note Digitizing is the process of splitting up a movement sequence into a large number of small incremental steps, and then defining the x-, $-y$ and z-co-ordinates of each incremental step's end point. Curved motion can be closely approximated by a series of linear movements providing a sufficiently large number of points are used (see also §9.2.2.3d).

Four of the commonest forms of IR used in manufacturing industry are shown in Figure 10.8.

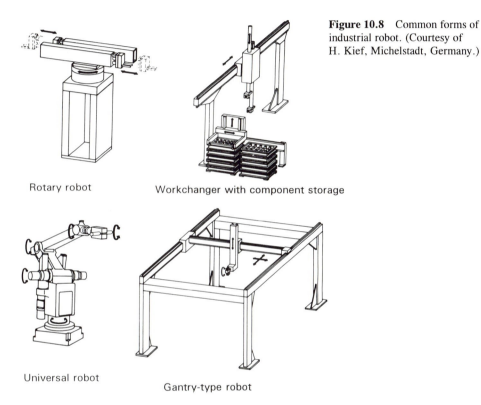

Figure 10.8 Common forms of industrial robot. (Courtesy of H. Kief, Michelstadt, Germany.)

Rotary robot Workchanger with component storage

Universal robot Gantry-type robot

Bibliography

Buffa, E. S & R. K. Sarin 1987. *Modern production/operations management*, 8th edn. Chichester: John Wiley.

Edwards, L & M. Endean 1990. *Manufacturing with materials*. Milton Keynes: Butterworth Scientific.

Mather, H. 1989. Simplification should precede automation. *Automation* **26**(4), 21–5.

Stanton, C. 1990. How will you make it? *Automation* **27**(3), 21–2.

Wild, R. 1989. *Production and operations management*. London: Cassell Educational.

Case study

The circular steel component illustrated in Figure 10.9 is to be produced initially in a prototype batch of ten. Thereafter, production quantities are estimated to be 10 000 per year, produced as required, in batches of 1000 each. What options are open to the production engineer when planning the manufacture of the first batch of 10, and thereafter the batches of 1000?

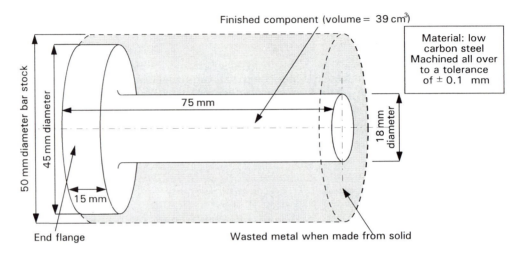

Figure 10.9   Component with end flange.

Solution

For the initial batch of ten components it is quicker and cheaper to machine them from standard 50 mm diameter bar, despite the fact that 108 cm³ of metal will be wasted (ignoring the metal lost in parting off the part from the 50 mm bar). For the larger batch quantity of 1000 the choice widens to include friction welding (§5.2.4.1) the end flange onto round bar, and possibly also impact extrusion (§3.5.3).

With the friction welding option it would still be necessary to finish machine the parts all over to meet the general ±0.1 mm tolerance requirement and because standard 50 mm and 20 mm diameter bar stock would be used. Cutting off appropriate lengths of both 50 mm and 20 mm bar would also be necessary.

In this particular example the costs of all these operations would almost certainly more than outweigh any savings resulting from improvements in material use – from 26.5 per cent use when machined from solid to 81 per cent when friction welding two parts together. Machining on a capstan lathe or CNC turning centre is therefore probably the preferred choice.

The impact extrusion option would use only 20 mm diameter bar stock with no 50 mm diameter bar being required. Machining all over is again required but, even allowing for this, material utilization would exceed 80 per cent – an excellent figure.

However, die costs could be high unless the design can be changed to suit the capabilities of an existing die. Furthermore, it would not be economic to impact extrude batches of just 1000, and either 6 months' or a year's production would probably be produced at one time. This incurs the costs of the material and labour involved in producing a high quantity in one batch, as well as having to store a larger than necessary stock level for much of the year. From the labour and number of operations standpoints, impact extrusion is more attractive than friction welding, but the need to set up each component for machining all over (two operations) means that machining from the solid is still probably the best option, despite the poor material utilization that this involves.

Questions

10.1 Define and differentiate between the following terms: (a) jobbing production, (b) batch production, (c) mass production. Describe the plant layout and work flow patterns for each of these three types of production. Consider specifically components that are produced on machine tools. Discuss the problems associated with work planning, control and labour requirements for each of the three types of production.

10.2 Mechanical equipment can be arranged in a number of ways, depending upon the form and quantity of product required. What are the most common configurations used and for what types of manufacturing are they best suited?

10.3 Describe what is meant by the term flow-line balancing.

10.4 What are the three prime elements in manufacturing that must be optimized if the most cost-effective production is to be realized?

10.5 Should the sole consideration of a designer be to design components that can be made with the minimum loss of material? If not, what other important factors should be taken into consideration?

10.6 Generally people need some form of incentive to work to best advantage. Other than financial, what incentives would cause you to produce your best work consistently?

10.7 Constantly improving productivity levels is the goal of all production managers. Suggest various ways that could be used to measure such improvements, indicating the relevance and limitations of each of your proposals.

10.8 What is the fundamental difference between automation and mechanization? Which is the more expensive to implement and why?

10.9 Describe the fundamental differences between fixed and programmable forms of automation. Where would each be best employed?

10.10 Discuss the statement "automate or liquidate".

10.11 What factors might lead a management team to decide not to automate their production? What are the possible outcomes of such a decision?

10.12 Does automation automatically imply lack of flexibility? If not, how is flexibility achieved in a machine tool manufacturing environment?

10.13 Before automating any process, what is the most important preliminary activity to ensure that the maximum benefit will be realized?

10.14 What is the single most important rule to be obeyed when deciding what to automate and what to leave to human effort?

10.15 Why does the elimination or reduction of set-up time greatly improve the productivity of small batch runs?

10.16 What market trends are driving companies towards small batch production?

Chapter Eleven

Quality assurance

After reading this chapter you should understand:

  (a) what is meant by quality

  (b) the benefits of international quality standards

  (c) the difference between quality assurance, quality control and inspection

  (d) the true costs of quality

  (e) the difference between conformance and accuracy

  (f) the relationship between process capability and tolerance

  (g) the importance of consistent process control and the value of statistical process control

  (h) the basis of length and angle measurement

  (i) how accurate measurements are performed

  (j) the difference between direct reading measurement and limit gauging

  (k) why the co-ordinate measuring machine is so important in modern manufacturing.

11.1 What is quality?

In a consumer society everyone is frequently a customer, whether the purchase involves a pen or a PC. While the costs of buying these two items are very different, the customer's expectations are precisely the same, namely that the product shall be of satisfactory quality and suitable for the purpose for which it has been sold. Indeed, the customer is protected by UK law [Sale of Goods Act 1979, Sections 14(2) and 14(3), Sale and Supply of Goods Act 1994, Section 1(1)] to ensure that this is so.

Therefore quality may be defined as the ability of a product or service to satisfy a stated or implied need (British Standards Institution 1993a), and should not be confused with reliability, which is the ability to maintain consistent quality. For manufacturers to be able to guarantee the customer consistent quality they must control the quality of all aspects of their business. Obviously, the more sophisticated the product the more complex will be the organization necessary to ensure that consistent quality is maintained.

To assist companies in setting up and maintaining suitable quality systems, in 1979 the British Standards Institution (BSI) introduced BS 5750:1979, which provided a series of specifications and guidelines for quality management and quality system elements. In 1987 the International Organization for Standardization (ISO) issued a specification on the same topic area, ISO 9000. BSI then released a harmonized version, BS 5750 (British Standards Institution

1987), and in the same year the European Union published its equivalent specification, EN 29000. From that time there has been international uniformity of the standards specifying good practices in the organization and management of quality assurance. The most recent revision of this international standard was released in 1994, and in the UK is known as BS EN ISO 9000 (British Standards Institution 1994).

11.2 Quality standard BS EN ISO 9000:1994

To consider the above unified international standard as being applicable only to manufacturing industry would be incorrect as its basic principles and aims can be applied with equal benefit to service organizations as diverse as higher education and the hotel industry. In every situation it should result in marked and sustained improvement in the product offered to the customer. Indeed, gaining certification to ISO 9000 is rapidly becoming an essential goal for companies in both manufacturing and service industries.

BS EN ISO 9000 lays down guidelines of how to establish and operate an efficient quality assurance system, covering most aspects of a business and its procedures, and specifies that such procedures must be documented in a quality assurance manual.

How the system is implemented, managed and periodically reviewed to ensure compliance and continued effectiveness also has to be clearly documented. Thus customers of any ISO 9000-approved company should feel reassured that they are buying from an organization that exercises tight control over its whole business, and that the end product will be consistently to the declared specification. As a result, many companies, both large and small, most government agencies and public service industries are becoming increasingly reluctant to deal with any supplier who does not meet the requirements of ISO 9000.

Exercise   Look in a trade directory for your area (e.g. Yellow Pages) and select a small number of local engineering companies of differing size. Write to them and ask whether they have either BS 5750 or ISO 9000 approval, and if not whether they are actively working towards this objective.

11.3 Differences between quality assurance, quality control and inspection

Some organizations use the terms quality assurance (QA) and quality control (QC) as alternative terms to describe inspection (I). This is unfortunate as they are very different functions.

If a customer is to be certain of being supplied with a product that fully conforms to specification, the supplier must operate an internal monitoring system to guarantee that the goods are correct at the point of sale. This is the prime task of a company's QA organization, but it is only possible to achieve this if consistent quality is maintained at all stages of manufacture by effective QC.

Inspecting goods before despatch, while preventing (most) faulty items from being despatched, does nothing to prevent rejects being made in the first place. The resulting costs of rectification or replacement, and possible late delivery, reduce company profitability and mar its reputation with its customers.

Clearly a much better philosophy is to control the quality of every process. In this way each successive process will be supplied with parts that guarantee satisfaction up to that point, and if this is carried through the whole manufacturing cycle the quality of the end product is assured.

How QC is carried out to ensure that all processes conform to specification varies with both the process and the manufacturing volumes involved. However, there should be one common objective, namely to detect immediately any significant drift away from specification so that remedial action can be taken before items are produced outside specification limits. In other words, *prevention is better than cure*.

Despite inspection being an expensive activity, some inspection will always be necessary, although its primary objective should be to monitor the process for consistency rather than to find rejects. It is also a fact of life that inspection is never 100 per cent efficient, resulting in a few rejects escaping detection and other items being incorrectly rejected.

Thus it can be said that $QA \rightleftharpoons QC + I$.

11.4 Cost of quality

How the costs of achieving a given level of quality are assessed has been subject to major reappraisal in recent years, but to appreciate current thinking it is first necessary to understand the terminology generally used when evaluating quality costs.

- The costs of activities designed to avoid defects, such as staff training programmes, increasing the number and/or sophistication of in-process automatic inspection devices and carrying out process capability assessments (§11.4.2), are all classed as prevention costs.
- The costs of activities such as post-operative inspection, the purpose of which is to detect whether defects have been made or not, are termed appraisal costs.
- The costs of all defects produced within the manufacturer's premises that have to be either rectified or replaced are referred to as internal failure costs.
- All expenses incurred as a result of in-service failures, such as warranty and other consequential claims are classed as external failure costs.

Prevention and appraisal costs are incurred in activities that are designed to prevent the production of scrap and to ensure that the product conforms to the required quality. The sum of the two are therefore referred to as conformance costs.

The expense of rectifying or replacing defective items, either before despatch or afterwards, because quality did not conform to the required standard are called non-conformance costs. These are the most difficult to quantify, as the cost of losing customer confidence, which in turn may well result in lost future orders, should also be included.

The overall cost of quality is the sum of both conformance and non-conformance costs, and BS 6143 (British Standards Institution 1992) suggests that they follow the form shown in Figure 11.1a. However, this leads to the conclusion that there is a cost region where quality costs are the most cost-effective. Unfortunately, the logical conclusion of this approach is that it is uneconomical to strive to make something with zero defects – clearly not an impressive approach from the customer's viewpoint!

Current thinking is that, while non-conformance costs may still follow the pattern suggested in Figure 11.1a, conformance costs do not. This shift of opinion is based upon the concept that, with effective quality controls at every stage of manufacture being a requirement of modern quality systems, *all* staff must be involved *all* the time to ensure that quality is maintained. This

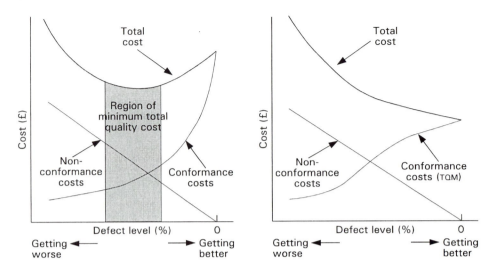

Figure 11.1   (a) Total cost of quality assuming ever increasing conformance costs. (b) Total cost of quality using the TQM approach.

is what is often referred to as total quality management (TQM). Once TQM is fully installed and operating effectively, there is no reason why conformance costs should continue to rise as suggested in Figure 11.1a. Indeed, total costs continue to decrease as the zero per cent defect level is approached because of reducing non-conformance costs (Fig. 11.1b).

Clearly this is a much more realistic conclusion, and aligns with the commonsense view that the aim should always be to get it right first time.

Exercise   Is reliability the same as quality? If not, what is it and how does it influence the customer's perception of quality?

11.4.1 Conformance and precision

It is the job of the manufacturing engineer to comply with specified conformance levels at the lowest possible cost, and not to redefine them by producing to an unnecessarily high level of precision. In other words, accuracy specification (tolerances) should always be made as large as possible while still meeting the conformance limits set for the end product. To do otherwise is to waste potential profit because, generally, the greater the level of precision demanded, the more expensive the process becomes (Fig. 11.2).

11.4.2 Process capability and tolerance

No component or product can be made consistently to an exact size because some random variation is inevitable in all manufacturing processes (and in every method of measurement). Therefore each dimension in an engineering specification must have a stated maximum allowable variation, and this is known as its tolerance.

The random scatter of a dimension often follows a Gaussian normal distribution pattern in

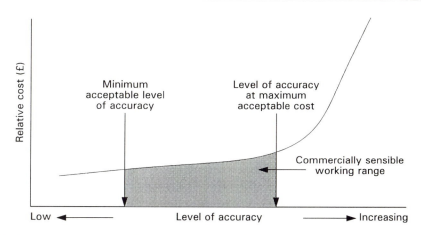

Figure 11.2 Typical cost curve applicable to most manufacturing processes.

which 95 per cent of all dimensions fall within a zone ±2σ from the mean and 99.73 per cent fall within a zone ±3σ from the mean. This scatter is termed the process capability (Fig. 11.3).

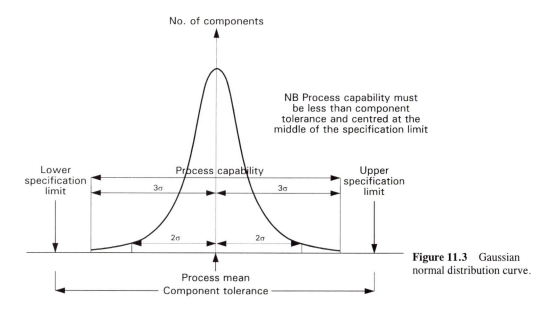

Figure 11.3 Gaussian normal distribution curve.

Note   It is assumed that the reader has a basic knowledge of simple statistics (Grant & Leavenworth 1988) and is aware that σ is the symbol used to represent standard deviation and that

$$\sigma = \sqrt{\frac{\sum \left(X - \overline{X}\right)^2}{N-1}}$$

where X is the value of individual sample readings, $\overline{X}$ is the arithmetic mean of all readings and N is the number of readings in sample.

A parameter called process capability ratio (C_p) is frequently used by industry and is a quantitative measure of component tolerance (T) with respect to process conformance (6σ) (for example, a machine tool's ability consistently to produce components within the required tolerance band), i.e.:

$$C_p = \frac{T}{6\sigma}$$

If rejectable dimensions are to be largely avoided then C_p must, of course, be at least equal to 1.0, but a C_p value of 1.33 is considered to be more realistic in practice. This indicates that the process is capable of producing 99.7 per cent of all parts within the centre 75 per cent ($1 \div 1.33$) of the tolerance band, and assumes that process scatter is centrally distributed about the middle specification limit (Fig. 11.3).

If the C_p value is lowered to 1.0 this increases the risk of rejects. Even so, statistically only three components per thousand would be outside tolerance limits, which in many situations is very acceptable; however, to the three unlucky customers the reject rate, as far as they are concerned, is 100 per cent. Furthermore, a C_p value of 1.0 gives no safety margin for process drift, for example because of tool wear.

11.4.3 Process control

It was suggested in §11.3 that it makes more commercial sense to use process control as an effective preventive measure than to rely solely upon post-process inspection to weed out defects. But to control most processes it is still necessary to inspect output periodically to ensure that the process is not slowly drifting towards an out of tolerance condition. If it is, then advance warning enables the operator to make the necessary adjustments before rejects start being produced.

In large-quantity production, if the manufacturing process is known to be stable, it is only necessary to inspect periodically a small sample of the process's output and, by applying some simple control rules, obtain a regular picture of process consistency over the manufacturing period, which can then be used to regulate the process efficiently. This approach is called statistical process control (SPC).

Exercise   SPC can also be used in certain aspects of mass production of high-value items such as motor vehicles. Make a list of areas where you would be happy to know that SPC had been used during the manufacture of your new car. Make a second list of areas where you would be most concerned by the use of this method of quality control.

While the arithmetic mean value ($\overline{X}$) of a series of readings is used to control the centrality of the process scatter, it gives no indication of the amount of scatter within a given process sample.

The difference between the maximum and minimum values within a sample (the scatter) is termed *range* (R). Although R is not as comprehensive or rigorous an indicator of sample scatter as the sample's *standard deviation* (σ), which takes account of *every* reading in the sample with respect to the sample's arithmetic mean value, it is much easier and quicker to calculate repeatedly on the shop floor when processing sample data manually, and it appears to be just as effective. However, with the increasing use of computer "on-line" processing of SPC data this objection is rapidly disappearing.

To depict graphically both average and range of each sample two charts are plotted, one directly below the other, using a common time base (Fig. 11.4).

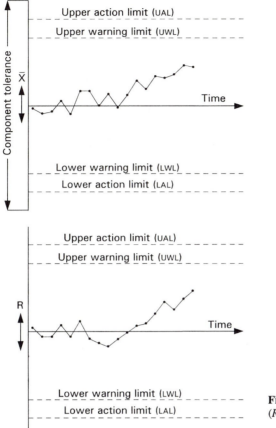

Figure 11.4 Arithmetic mean ($\overline{X}$) and range (R) charts.

The upper and lower *action* limits (UAL and LAL) on $\overline{X}$ charts are usually set at ±3σ for 99.7 per cent confidence of process control:

$$\text{UAL} = \text{target dimension} + \left(\frac{3\sigma}{\sqrt{N}} \right)$$

$$\text{LAL} = \text{target dimension} - \left(\frac{3\sigma}{\sqrt{N}} \right)$$

where N is the number of sample readings.

The upper and lower *warning* limits (UWL and LWL) on $\overline{X}$ charts are usually set at ±2σ for 95 per cent confidence of process control:

$$\text{UWL} = \text{target dimension} + \left(\frac{2\sigma}{\sqrt{N}} \right)$$

$$\text{LWL} = \text{target dimension} - \left(\frac{2\sigma}{\sqrt{N}} \right)$$

293

Target dimension can be either the middle of the specification limits, the current process mean or a customer-specified value.

Sample range, R, charts also have action and warning limits. These are calculated by multiplying the sample's standard deviation (σ) by a factor from tables found in publications such as BS 2564 (British Standards Institution 1993b) (Table 11.1). Commonly a sample size of four or five is used to minimize sampling time.

Table 11.1

Sample size	Upper action	Upper warning	Sample mean	Lower warning	Lower action
N	$D_{0.001}$	$D_{0.025}$	D_N	$D_{0.975}$	$D_{0.999}$
4	5.30	3.98	2.059	0.59	0.20
5	5.45	4.20	2.326	0.85	0.37

Upper *action* (control) limit $= D_{0.001} \times \sigma$; Upper *warning* limit $= D_{0.025} \times \sigma$; Mean range $R = D_N \times \sigma$; Lower *warning* limit $= D_{0.975} \times \sigma$; Lower *action* (control) limit $= D_{0.999} \times \sigma$.

A practical example illustrating how $\overline{X}$ and R action and warning limits are calculated is given in the case study at the end of this chapter.

Commonly used rules to decide when to stop the process and to take action to restore the "in control" state are when:
- any *one* point is plotted outside an action limit
- *two* out of five successive points are plotted outside a warning limit
- any *seven* successive points show an increase from the previous point
- any *seven* successive points show a decrease from the previous point
- any *seven* successive points are above the mean
- any *seven* successive points are below the mean.

11.5 Measurement of length and angle

The ability to control a manufacturing process presupposes the ability to measure reliably the features of products produced. Two of the most common features measured in engineering are length and angle, and the precision to which these parameters must be measured is a function of the specified limits to which the product has to be made. It is also influenced by the component's physical size. One could hardly anticipate needing to measure a 5 m long feature to an accuracy of, say, 2 µm, but it may well be necessary to measure a 10 mm dimension to this accuracy.

For measurements to have credibility, they must be obtained using procedures and equipment of proven accuracy that can be traced back to national and international standards via a system known as calibration.

A simple calibration rule is that the accuracy of the measuring equipment should be ten times as great as the allowable tolerance of the component being measured. This rule is applied all the way back along the measurement chain from the product being measured, through working standards such as micrometers (§11.5.2), laboratory standards such as gauge blocks (§11.5.1), national standards in the form of calibrated reference gauge blocks, to international standards (Fig. 11.5). Thus, for a product made to a tolerance of ±0.01 mm the international standard of length must be known to ±0.000001 mm (±1 nanometre!).

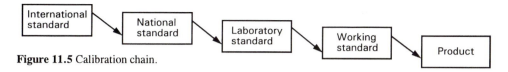

Figure 11.5 Calibration chain.

11.5.1 Reference standards for length and angle

The international standard of length is the metre. While its definition (Galyer & Shotbolt 1990), in terms of the velocity of light, is of academic interest only, it is essential that a company's measuring equipment can be traced back to this standard with an appropriate level of confidence.

The reference standards of length used by the engineering industry are precision gauge blocks – also known as slip gauges. They consist of a set of extremely accurate rectangular steel, carbide or ceramic blocks that are so highly polished on their ends that it is possible to wring them temporarily together (see Note below) to produce any required length (Fig. 11.6a) from 1.000 mm to over 250 mm in incremental steps of 0.005 mm (5 μm). Circular gauge bars are available for longer lengths, although the standard box of 88 gauge blocks (Fig. 11.6b) is usually adequate for most manufacturing needs.

(a)

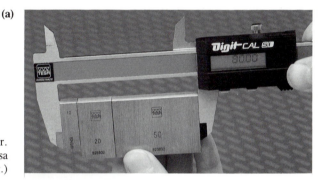

Figure 11.6 (a) Gauge blocks wrung together. (b) Box of gauge blocks. (Courtesy of Tesa Metrology Ltd, Leicester.)

(b)

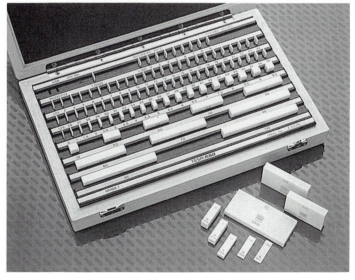

Note   When two extremely smooth and flat surfaces are pressed together, even under hand pressure, the two mating surfaces bond together as a result of molecular attraction. This is called *wringing*. With such highly polished surfaces the mating faces must be free of even the smallest dust particles or fingerprints if surface damage is to be avoided.

Warning Gauge blocks that are wrung together should not be left in this state for too long, as molecular attraction increases with time and separation then becomes difficult.

The reader should appreciate that *gauge blocks measure nothing*, and are principally used as reference lengths to check the accuracy of measuring equipment. The uncertainty of the sizes of grade 1 (inspection) gauge blocks should ideally be ten times less than the uncertainty of the equipment they are used to check (see calibration rule, §11.5). These gauge blocks should be periodically checked against master blocks that have been individually calibrated by an independent calibration laboratory to an uncertainty level of ±0.00012 mm (0.12 μm).

Working at such small levels of uncertainty is pointless other than in a temperature and hu-

(a)

Figure 11.7   (a) Angle gauge blocks wrung together.
(b) Box of angle gauge blocks.
(Courtesy of Tesa Metrology Ltd, Leicester.)

(b)

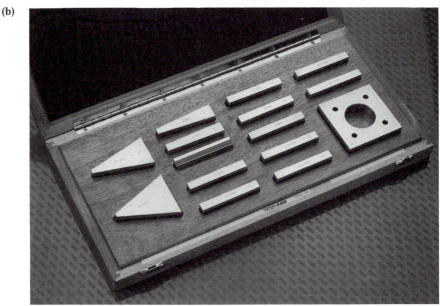

midity-controlled environment, the standard reference temperature being 20°C.

The unit of angular measurement is the degree, and is defined as 1/360 of a circle. Gauge blocks of exactly the same format as used to create linear reference lengths are employed to set up reference angles except that, instead of being parallel blocks, they are superprecision angle blocks (Fig. 11.7a). A complete set (Fig. 11.7b) offers a range of angle blocks which, when wrung together, can create any desired angle from 0° to 360° in increments of 3 seconds of arc. Unlike length gauge blocks, which can only be added together, angle gauge blocks can be combined either in addition or in subtraction. Thus, the 3° and the 1° blocks can be added to produce a 4° standard or the direction of one gauge block can be reversed to define a 2° angle.

Exercise   1° and 3° are the first two angle gauge blocks in a standard set. Show that with the addition of just the 9° and 27° angle blocks any whole angle up to 40° can be defined.

11.5.2 Basic direct reading measuring equipment

The majority of precision length and angle measurements made in engineering workshops and inspection departments are carried out using a relatively small number of basic measuring devices. While most are both simple and straightforward to use, they are nevertheless capable of giving reliable readings, providing they are used correctly and are regularly checked against inspection gauge blocks (§11.5.1).

For length measurement the most popular devices are the micrometer and the vernier calliper (Fig. 11.8). Modern digital display micrometers offer an accuracy of ± 0.003 mm and a resolution (smallest measuring increment) of 0.001 mm (0–25 mm working range), and a digital display vernier with a 0–150 mm working range and a resolution of 0.01 mm is accurate to ± 0.02 mm.

Exercise   One length-measuring device that is even more common on the shop floor than the micrometer or vernier is the engraved steel rule. While not a precision instrument, how accurately do you think you could measure with one of them? Measure as accurately as you can the length or diameter of a variety of small machined parts using a steel rule. Now measure them using a micrometer or vernier. With good eyesight you should find that your steel rule readings are accurate to better than 0.30 mm.

Micrometers and verniers are used for external, internal and depth/height measurement (Fig. 11.8), although micrometers are generally the more popular of the two because of their greater accuracy.

Angles are much more difficult to measure accurately by direct measurement, and the best piece of equipment that is usually available on the shop floor is the precision bevel protractor (Fig. 11.9a). Fortunately, the need for precise angular measurement is less common than for accurate linear measurement, but when high accuracy is essential either a sine bar (Fig. 11.9b) or special optical equipment such as an autocollimator or angle dekkor (Galyer & Shotbolt 1990) is used. Unfortunately, angular measurement using this type of metrology equipment involves removal of the part from the machine tool. It may then be impossible to reload the component in the machine accurately if minor corrections are found to be necessary.

A general observation applicable to most precision measuring equipment is that the smaller its working range the greater is likely to be its accuracy and resolution.

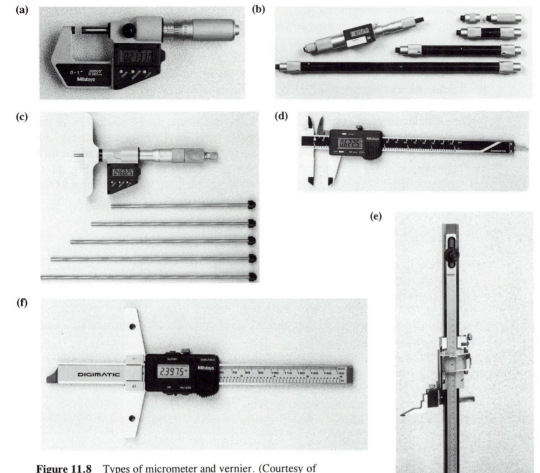

Figure 11.8 Types of micrometer and vernier. (Courtesy of Mitutoyo Ltd, Andover.)

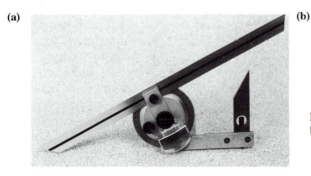

Figure 11.9 (a) Bevel protractor. (b) Sine bar. (Courtesy of Mitutoyo Ltd, Andover.)

11.5.3 Limit gauges and comparators

When manufacturing a significant number of identical parts, measuring critical dimensions of each one and then comparing the readings with the tolerances specified on the drawing is both time-consuming and prone to human error. However, it may be only necessary to confirm that the dimensions checked lie within their allowable tolerance bands. If they do, it may not matter what the exact dimensions are, although such an approach does rely heavily on the ability of the checker to detect, by feel, when parts start to approach one or other of the specified limits.

To confirm or otherwise dimensional compliance without actually measuring the part concerned is achieved by the use of limit gauges. These are simple devices of various forms used to check specific component features such as internal and external diameters (Fig. 11.10), thread size and form, tapers, etc. They all have one thing in common – they each have a GO and a NO GO portion. Normally the NO GO portion is much shorter than the GO portion because, if the part is within tolerance, the NO GO portion of the gauge should not pass over or enter the workpiece anyway.

Consider, for example, measuring the diameter of a shaft with a gap gauge whose GO dimension is set by gauge blocks to the maximum acceptable diameter (Fig. 11.11a). The NO GO gap is similarly set, but to the smallest acceptable diameter (Fig. 11.11b). Thus, when the gauge is placed on the diameter to be checked, for the diameter to be in tolerance the GO portion must pass over it (Fig. 11.11c); if it will not, the shaft is too big. The NO GO portion should not pass over the shaft because, if it does, the diameter has been made too small.

Such checking gauges are quick, efficient and generally more robust than conventional measuring instruments, although they are subjective in that they rely, to some degree, upon the feel of the user. This is particularly true when working to small tolerances.

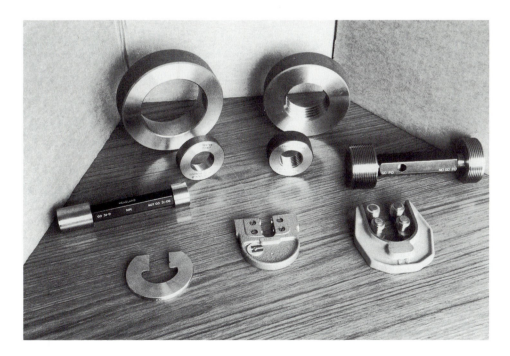

Figure 11.10 Selection of plug and gap gauges. (Courtesy of Coventry Gauge Ltd, Poole.)

(a)

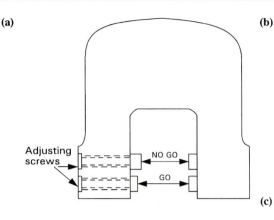

Adjusting screws

NO GO

GO

(b)

(c)

Figure 11.11 (a) GO/NO GO gap gauge. (b) NO GO portion of adjustable gap gauge being set. (c) Checking component diameter with a gap gauge. (Courtesy of Coventry Gauge Ltd, Poole.)

Plug gauges (Fig. 11.10), are used to check bore sizes (internal hole diameters). The GO diameter must coincide with the smallest acceptable bore size and the NO GO diameter to the maximum bore dimension.

Exercise Can you see why the limit gauging approach is unsuitable when using SPC? Could a gap gauge be used to check a shaft's circularity or concentricity about its rotational axis? If not, how would you check these important parameters?

Plug gauges must be specially made for holes with a specific tolerance specification and when, after much use, the GO end becomes worn, it has to be either replaced or reclaimed by chrome plating (§6.3.2.2) and reground back to the correct size.

Gap gauges are often adjustable over a small working range, gauge blocks being used to set the precise gaps (Fig. 11.11b), although non-adjustable gauges (called plate gauges), are also used. However, they are subject to wear on their GO end in the same way as plug gauges.

Other forms of GO/NO GO gauge that rely on optical, electronic and pneumatic principles (Galyer & Shotbolt 1990) are also available. Although these largely overcome the problem of operator feel, they are usually much more expensive, and generally cannot be used while the component is still in the machine.

Plug gauges are periodically checked for wear using a measuring instrument called a comparator (Fig. 11.12).

Comparators are not direct reading devices but, as the name suggests, they only compare the dimension being checked against a reference dimension (usually gauge blocks). They then display any difference between the two (Fig. 11.12b). Because comparators are required to dis-

(a)

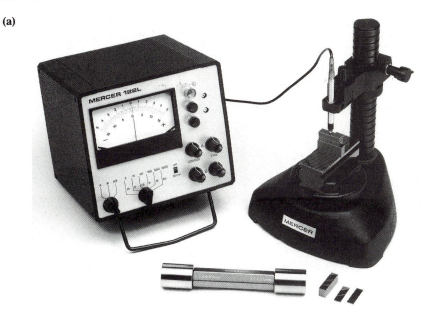

(b)

Deviation from gauge block height

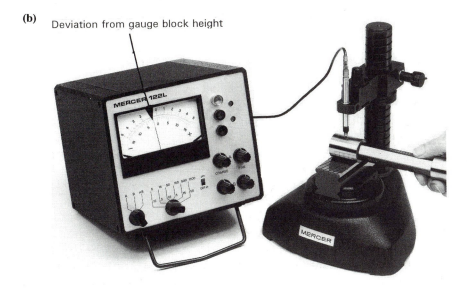

Figure 11.12 (a) Comparator being set up with gauge blocks to check plug gauge. (b) Comparator being used to check plug gauge accuracy. (Courtesy of Thomas Mercer Ltd, St. Albans.)

play (hopefully) small deviations from the reference dimension, their working range is very restricted – typically ±0.05 mm. To make extremely small deviations readily visible the working range is greatly amplified by a combination of mechanical and either optical or electronic methods. Typical maximum resolution is 0.0005 mm (½ μm), and even this can be subdivided by eye to ¼ μm.

Exercise   Take a 75 mm or 100 mm gauge block and set it up in a comparator. Carefully remove the gauge block and, without putting your fingers on either of the two "gauging faces", hold it is your hand for 2 or 3 minutes. Put the gauge block back in the comparator and notice how much it has expanded as a result of warming in your hand. This exercise also demonstrates the importance of the 20°C reference temperature at which all precision measuring equipment is calibrated, and the dangers of putting small gauges and micrometers in one's pocket or resting them on warm parts of a machine tool.

11.5.4 In-process gauging

Even if a component feature is measured while the part is still in the machine tool, if it is found that too much material has been removed, the part is scrapped. However, if the feature in question could be continuously monitored while cutting is proceeding, it would then be a simple matter to cease cutting when the desired dimension is reached, and so avoid the production of scrap.

Exercise   Can you see the similarity between this approach and the closed-loop control used in CNC machine tools? If not reread §9.2.3.1.

Unfortunately, the cutting zone of most machine tools is extremely hostile in that hot swarf is flying in all directions, the workpiece or cutting tool is moving, probably causing the workpiece to vibrate slightly, and the whole area is flooded with a high-pressure coolant supply. This is therefore hardly the place to site precision measuring equipment, and for this reason it has not yet been possible to produce commercially general-purpose continuous monitoring equipment capable of working under such arduous conditions.

However, there is one machining activity that does not generally involve large swarf removal rates, and that is grinding (§8.2.5.1). Because cutting conditions in grinding are therefore much less severe, continuous monitoring equipment is available (Fig. 11.13). This is fortunate, as grinding is frequently the final operation and usually involves working to extremely small dimensional tolerances. As the part also usually attains its maximum added value at this point, anything that reduces the risk of scrap is commercially very attractive.

Continuous in-process gauging should not be confused with what is generally termed in-cycle gauging (§9.2.2.6), as the latter refers to periodically stopping the workpiece, usually at the end of a cut, and then checking it before proceeding with further machining. This is a form of intermittent inspection, and is technically much easier to achieve as the part is stationary and not flooded with coolant when measurements are being taken.

While in-cycle gauging is not as comprehensive as continuous in-process monitoring, it is still extremely useful and greatly reduces the level of scrap associated with processes that rely solely upon post-process inspection.

Figure 11.13 Grinding machine component continuous monitoring device. (Courtesy of Marposs Ltd, Coventry.)

11.6 Co-ordinate measuring machines

The installation of CNC machine tools invariably leads to the rapid realization that existing QC and inspection methods are unable to cope with the wide range of complex components typical of CNC production, or to meet the exacting measurement standards required. Indeed, many companies find that their CNC machine tools are more accurate than any of their existing inspection equipment! Invariably the purchase of a co-ordinate measuring machine (CMM) is the solution to their problem.

11.6.1 What is a CMM?

The introduction of the CMM is one of the most significant advances in metrology equipment in recent years, and it is therefore essential that the reader gains a basic understanding of its construction and operation.

In its simplest form a CMM is a rigid structure designed to measure points on a component in all three axes simultaneously, outputting the resulting cartesian (or polar) co-ordinates for each point measured. As the output is in the form of an electronic signal, readings can be fed directly into a computer which, with suitable customized software, can be used to compute the size and location of a wide range of features such as lines, planes, holes, tapers, etc. From these computations parameters such as straightness, squareness, roundness and concentricity can be calculated at great speed and devoid of human error.

Both manually operated and CNC CMMs are available (Fig. 11.14), the latter being program-

303

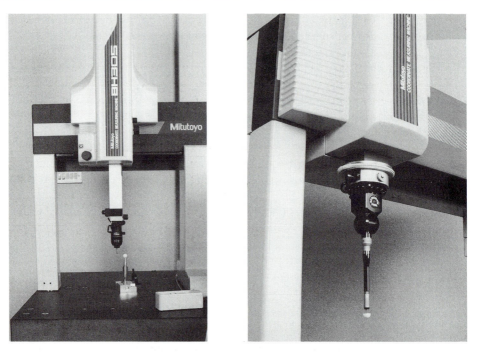

Figure 11.14 Typical manual and CNC types of CMM. (Courtesy of Mitutoyo Ltd, Andover.)

mable and therefore able to check automatically all the required features for a whole batch of parts once the first one has been manually probed in "teach-in" mode (§10.9.2).

To ensure reference traceability back to national standards (as required by BS EN ISO 9000), CMMs must be calibrated at regular intervals; most manufacturers offer this service.

11.6.1.1 Structure

There are many configurations of CMM, depending upon the size and shape of the parts to be measured, but the most common one is the bridge type shown in Figure 11.15.

Figure 11.15 Bridge-type CMM.

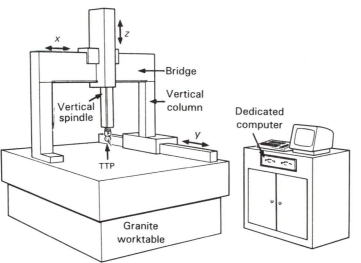

The worktable is usually made of granite because of its rigidity, great dimensional stability and hardness, although ceramics are also becoming increasingly popular owing to their reduced mass compared with granite.

The vertical columns and the bridge that they support used to be made of fully stress-relieved steel, but increasingly aluminium is being used because of its lower mass. This requires reduced traversing forces and makes faster axis traversing speeds possible. Measurement in the z-axis is achieved by traversing the vertical spindle unit, which is supported from the machine's horizontal bridge.

Slideways in the x-, y- and z-axes have to be as friction free as possible, and this is why air bearings are invariably used. Note that, in contrast to CNC machine tool builders, traditionally CMM manufacturers use the left-hand rule in determining the direction of the x- and y-axes (Fig. 11.15).

It is essential that the CMM sits on vibration-proof mountings and is housed within a temperature-controlled environment ($20°C \pm 1°C$) if the highest resolution levels and measurement accuracy are to be maintained.

CMM measurement accuracy may be expressed by the equation $(K + HL)\,\mu m$, where K and H are constants and L is the measuring length, in metres, over which the accuracy is being calculated (International Organization for Standardization 1993). Typical values of K and H are 4 and 5 respectively.

Exercise   Can you suggest why it is necessary to leave a component in the CMM environment for some time before attempting to take any measurements?
*Hint*s The larger its mass the longer it must be left before accurate measurements are attempted; the coefficient of linear expansion of steels is 0.000012 /°C.

Two of the most vital components of any CMM are the linear measuring scale fitted to each axis, and its measuring probe.

11.6.1.2 Axis measuring scales

There are a number of design requirement similarities for both CNC machine tools and CMMs, but perhaps the most obvious one is that extremely accurate length measurement in each axis is vital.

In Chapter 9 (§9.2.3.2) a brief description was given of how rotary encoders are used as the basis for linear measurement on most CNC machine tools. Although this method is also sufficiently accurate for the high levels of precision required of CMMs, a number of CMM manufacturers prefer to use a linear scale permanently attached to each axis. This scale is not a precision tape measure, but rather a strip of material consisting of a grid of alternate equispaced opaque and transparent strips, making it look rather like a bar code (Fig. 11.16). As the strip moves it is optically scanned and referenced against a fixed grid. This type of linear measuring scale is termed a diffraction grating, and provides extremely high resolution levels down to 0.1 μm.

Note   Diffraction grating systems are also used on some CNC machine tools, mainly from the Far East. They are inherently more accurate than rotary encoders, but to be used on CNC machines they have to be sealed against the ingress of swarf, dirt and coolant – not a problem on CMMs. Providing the accuracy of the encoder form of measuring system is better than the inherent accuracy of the CNC machine, there is little point in fitting an even more accurate axis-measuring system, particularly if it presents an environmental sealing problem.

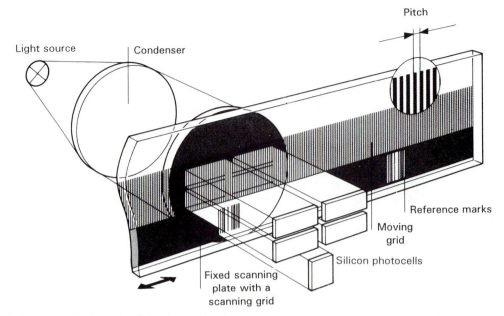

Figure 11.16 Operating principle of a diffraction grating linear measuring system. (Courtesy of Heidenhain (GB) Ltd, Burgess Hill.)

11.6.1.3 Touch trigger probes

The uses of TTPs on CNC machine tools were briefly discussed in section 9.2.2.6, but a detailed description of their operating principle has been postponed until now. This is because CMMs cannot work without a TTP whereas, if a CNC machine has a cutting tool, it will function, albeit less efficiently than when fitted with TTP facilities.

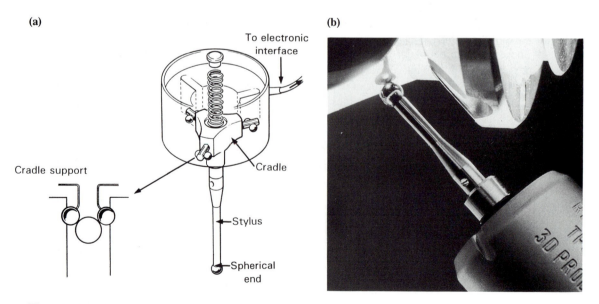

Figure 11.17 (a) TTP operating principle. (b) Typical commercial probe. (Courtesy of Renishaw plc, Wotton-under-Edge)

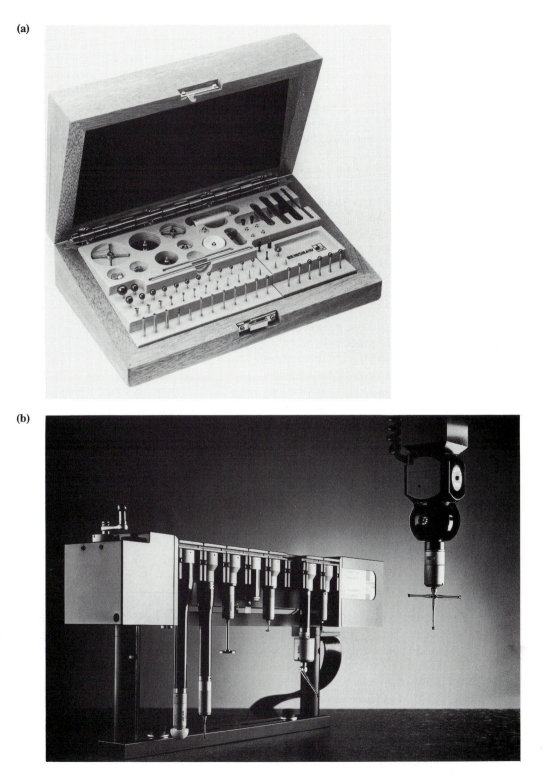

Figure 11.18 (a) Box of probe styli. (b) CMM probe autochanger unit. (Courtesy of Renishaw plc, Wotton-under-Edge.)

A TTP consists of a spherical-ended stylus rod (typically 50–75 mm long), which is supported in a spring-loaded three-sided cradle, and wired to form an electrically balanced bridge (Fig. 11.17a). When the end of the probe touches the workpiece the cradle is momentarily unseated from at least one of its three cradle supports, immediately putting the bridge circuit out of electrical balance. This then generates a signal that is used both to stop the probe traversing further and to trigger a measurement of the probe's position in free space at that instant.

Linear position is measured by reference to the precision diffraction grating (§11.6.1.2) fitted to each of the CMM's axes.

A wide range of probe sizes and configurations are available (Fig.11.18a) to meet the demands of measuring the most complex component features. Even an automatic probe changer, similar to automatic tool chargers on CNC machine tools, is available for use on CNC CMMs (Fig. 11.18b).

It is important to realize that, because TTPs are no more than highly reliable on/off switches, *they cannot have a measuring accuracy*, only a repeatability accuracy – typically ±1 μm.

11.6.2 Making measurements on a CMM

There are five basic steps involved in inspecting a component using a CMM. These are:
1. Calibrate the probe system.
2. Define component datum surfaces.
3. Perform the actual measurements.
4. Calculate the required dimensions using the measurements taken in step 3.
5. Compare calculated dimensions with specification.

11.6.2.1 Calibration of the probe system
A TTP has a spherical end, and it is the surface of this sphere that touches the component when taking measurements (Fig. 11.19).

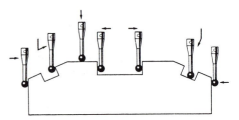

Figure 11.19   Feature probing. (Courtesy of Renishaw plc, Wotton-under-Edge.)

Because CMMs make measurements with respect to the centre of the TTP's sphere, the sphere must be calibrated to establish its precise diameter and geometric centre. This is achieved by reference to a calibrated steel sphere mounted on the granite work surface of the CMM. It also establishes the centre of the TTP's spherical end with respect to the CMM's x, y and $z = 0$ datum points (set by the CMM builder at the time of commissioning). Each time a measurement is made, the CMM software automatically applies the appropriate x, y and z offsets to compensate for the difference between the probe's spherical surface in contact with the workpiece and its geometric centre.

Exercise   Can you see any similarities between CMM probe and CNC machine cutting tool offsets? If not reread §9.2.2.4 and §9.2.2.6.

11.6.2.2 Definition of component datum surfaces

When component features are measured, the dimensions taken must be relative to the component's datum points shown on its drawing, and not relative to the CMM's x, y and $z = 0$ datums. Therefore, because positioning of a component on the worktable of a CMM is completely arbitrary, before any meaningful measurements can be made the part's datum surfaces must be defined relative to the CMM's axis measuring system.

Any rigid body in free space has six degrees of freedom; thus, full definition of the position of a workpiece with respect to the CMM's measuring system requires a minimum of six touch points over its surface (Fig. 11.20). The minimum number of touch points to define common geometrical features (Fig. 11.21) is: line, 2; plane, 3; circle, 3; sphere, 4; cylinder, 5; ellipse, 5; cone, 6.

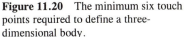

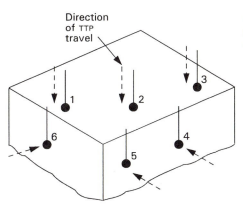

Figure 11.20   The minimum six touch points required to define a three-dimensional body.

Most CMM computer software offers operator prompting to ensure that at least the minimum required number of touch points necessary to measure a given feature are actually made.

11.6.2.3 Performing the actual measurements

The process of touching a component at the various points necessary to calculate required feature dimensions is carried out manually on small CMMs, but on CNC CMMs the TPP can be driven either by a joystick control or by component preprogramming.

While the minimum number of touch points must be used when checking a given feature (Fig. 11.21), generally the more points taken the more confidence can be placed on the resulting calculated dimension. While this increases probing time slightly on manually operated machines, with CNC machines operated in automatic probing mode the difference in overall measurement time is minimal.

11.6.2.4 Calculation of dimensions

The measurement points taken in step 3 are the figures used to calculate the various feature dimensions required.

Calculations can be quite complex, as probe offsets and allowances for the difference between component and CMM datum planes all have to be taken into account. Fortunately, the software supplied with all CMMs carries out these calculations automatically, making life simple for the operator, as well as eliminating the ever-present risk of human error associated with manual calculations.

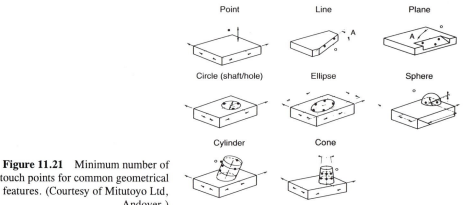

Figure 11.21 Minimum number of touch points for common geometrical features. (Courtesy of Mitutoyo Ltd, Andover.)

11.6.2.5 Conformance to specification

The dimensions calculated in step 4 must finally be compared with the component's specification, and an assessment made of whether each feature examined is within acceptable limits. It is normally possible to print out the calculated results as a permanent record if required.

Other analyses of results can also be readily performed on the CMM's computer and are limited only by the capabilities of the software employed.

11.6.3 Summary of the advantages of CMMs

11.6.3.1 Versatility

A CMM is capable of performing a wide range of measurements, some of the most common examples being length, external and internal diameter, position, three-dimensional angles and solid shape geometries such as cylinders, spheres and cones (Fig. 11.21). Thus, a CMM can be looked upon as a precision *measuring centre*.

Readings can be used as the basis of accept/reject inspection, as well as providing data within a closed-loop feedback system for process control such as monitoring cutting tool wear and detecting part positioning errors on CNC machine tools.

CMMs can also be used to measure prototype models, dies, etc., the resulting data then being used to create drawings on a CAD system and/or to provide geometrical data for CNC machine tool programs.

11.6.3.2 Improved accuracy

A single set of measurement datums from which all readings are measured eliminates the accumulation of errors that often occurs if parts have to be repositioned to measure related features – the usual situation when measuring manually on a conventional inspection table.

The contact force used by the TTP (typically 3–8 g), causes no job movement or distortion, even on lightweight components.

11.6.3.3 Operator influence minimized

TTPs eliminate operator feel, and human error is avoided as all calculations are performed by the computer. Large volumes of measurement data can be stored in the computer and/or on floppy disk, and printed reports in the required format are readily available if required.

11.6.3.4 Reduced measurement set-up times

A single set-up can be used for the measurement of the majority of dimensions on a workpiece. Special fixtures are seldom required as alignment of component's axes to the CMM's measuring system is automatically accounted for by the CMM's computer software. Also, clamping the component to the work surface is rarely necessary because of the TTP's low contact force.

11.6.3.5 Improved productivity

Greatly reduced inspection times minimize delays to the whole manufacturing system, and early warning of dimensional errors or changes can help avoid the manufacture of rejects.

As CMM technology is compatible with computer-aided engineering (CAE) methodologies, its integration into a total manufacturing system is possible. Indeed, the CMM is now becoming a vital element in many automated and flexible manufacturing installations.

Bibliography

British Standards Institution 1987. *Quality systems, BS 5750*. London: British Standards Institution.

British Standards Institution 1992. *Guide to the economics of quality*, Part 1, *Process Cost Model, BS 6143*. London: British Standards Institution.

British Standards Institution 1993a. *Quality vocabulary, BS 4778*. London: British Standards Institution

British Standards Institution 1993b. *Control chart technique when manufacturing to a specification, with special reference to articles machined to dimensional tolerances, BS 2564*. London: British Standards Institution.

British Standards Institution 1994. *Quality Management and Quality Assurance Standards, BS EN ISO 9000*. London: British Standards Institution.

Galyer, J. F. W. & C. R. Shotbolt 1990. *Metrology for engineers*, 5th edn. London: Cassell.

Grant, E. L. & R. S. Leavenworth 1988. *Statistical quality control*, 6th edn. New York: McGraw-Hill.

International Organization for Standardization 1993. *Performance verification of co-ordinate measuring machines, 10360–2*. London: British Standards Institution.

Case study

A. A process is making a component with a dimension specified as 30 mm ± 0.08 mm. It is known that the process capability ratio $C_p = 1$ and that, by adjustment, the process can be centred onto any desired value. A sample size of four is used. Using the data contained in Table 11.1 of this chapter verify that limit lines for the $\overline{X}$ and R charts will be as shown in Table 11.2.

Table 11.2

Limit	Sample average	Sample range
Upper action limit	30.040	0.142
Upper warning limit	30.027	0.106
Mean	30.0000	0.055
Lower warning limit	29.973	0.016
Lower action limit	29.960	0.005

Solution

With $C_p = 1$ and tolerance $= 0.16\,\text{mm}$ then $\sigma = 0.16/6 = 0.0267\,\text{mm}$.
Sample number $(N) = 4$. Target value $= 30\,\text{mm}$.

Limits for average chart

$$\text{UAL} \quad 30 + \frac{3\sigma}{\sqrt{N}} = 30 + 0.04 = 30.04\,\text{mm}$$

$$\text{LAL} \quad 30 - \frac{3\sigma}{\sqrt{N}} = 30 - 0.04 = 29.96\,\text{mm}$$

$$\text{UWL} \quad 30 + \frac{2\sigma}{\sqrt{N}} = 30 + 0.027 = 30.027\,\text{mm}$$

$$\text{LWL} \quad 30 - \frac{2\sigma}{\sqrt{N}} = 30 - 0.027 = 29.973\,\text{mm}$$

Limits for range chart

UAL	$D_{0.001} \times \sigma$	$= 5.30$ (From Table 11.1) $\times\ 0.0267$	$= 0.142\,\text{mm}$
LAL	$D_{0.999} \times \sigma$	$= 0.2 \times 0.0267$	$= 0.005\,\text{mm}$
UWL	$D_{0.025} \times \sigma$	$= 3.98 \times 0.0267$	$= 0.106\,\text{mm}$
LWL	$D_{0.975} \times \sigma$	$= 0.59 \times 0.0267$	$= 0.016\,\text{mm}$
Mean	$D_N \times \sigma$	$= 2.059 \times 0.0267$	$= 0.055\,\text{mm}$

B. Table 11.3 shows the measured sizes of four components in each of five samples taken when controlling the manufacture of the product described above. Calculate the average and range figures and comment on the state of control shown by each sample. *Note than none of the measured dimensions is outside the component tolerance of ±0.08 mm.*

Table 11.3

Sample	1	2	3	4	5
Component 1	30.05	30.07	30.03	30.06	29.95
Component 2	30.02	30.01	30.02	30.05	29.95
Component 3	30.00	29.98	29.98	30.05	29.95
Component 4	29.95	29.92	29.97	30.02	29.95

Solution

Table 11.4

Sample	1	2	3	4	5
Sum	120.02	119.98	120.00	120.18	119.80
Average	30.005	29.995	30.00	30.475	29.95
Range	0.10	0.15	0.06	0.04	0.00
Comments	In control	Range is above UAL -scatter is too high	In control	Average is above UAL -setting is too high	Average is below LAL *and range is too low*!

Questions

11.1 Explain what is meant by the word quality.

11.2 Why are an increasingly wide range of commercial businesses working to the international standard specification for quality systems?

11.3 Discuss the statement "quality costs money". Take into account the fact that there is a limit to what the customer is willing to pay for a product of given specification and quality.

11.4 What are conformance and non-conformance costs and how are they affected by the "get it right first time" philosophy of TQM?

11.5 Why is there a need for quality assurance in modern industry, and how does it differ from both quality control and inspection?

11.6 Explain what is meant by process capability, and explain how it is taken into account when planning the manufacture of a particular component.

11.7 SPC is only of value in certain forms of manufacturing. Why is it not possible to use it universally?

11.8 Why is standard deviation preferred to range as the parameter for assessing process scatter? Despite this, why do many of today's SPC practitioners still prefer to use range?

11.9 Define the SI standards of length and angle.

11.10 What does manufacturing industry employ as its length and angle reference standards, and how may they be related to national and international standards?

11.11 What is the fundamental difference between a direct and an indirect measuring system? Are comparators direct reading devices?

11.12 How can the NO GO portion of a limit gauge be easily recognized? What is the biggest single limitation of a plug gauge but which does not apply to a gap gauge?

11.13 Briefly describe the difference between continuous in-process gauging and in-cycle gauging.

11.14 Compare the difference between dimensional quality control of a machined component by in-process and post-process gauging.

11.15 Describe the essential constructional features of a CMM.

11.16 What are the principal advantages of using a CMM for inspecting workpieces compared with using conventional measuring instruments such as micrometers, surface plates, height gauges, dial indicators and gauge blocks?

11.17 Explain the principle of measurement using diffraction gratings. How is this applied to ensure precise linear axis measurement on CMM slideways and why is it less attractive as a means of axis measurement on CNC machine tools?

11.18 How does a TTP interact with a CMM? Briefly describe its two prime functions.

11.19 What facilities would you expect to find in a comprehensive CMM software package?

11.20 It is proposed to incorporate a CMM into an automated manufacturing cell. What uses could be made of the measuring data it generates, and what other hardware facility will probably be required in order to fully integrate the CMM into the cell?

Index